Current Topics in Neuroendocrinology Volume 5

Actions of Progesterone on the Brain

Editors
D. Ganten and D. Pfaff

Contributors
J. Kato L. Kiesel B. Parsons D. Pfaff Th. Rabe
J. M. Reinisch B. Runnebaum Y. Sakuma
S. A. Sanders P. Södersten

With 61 Figures

Springer-Verlag
Berlin Heidelberg New York Tokyo

Editors

Dr. DETLEV GANTEN, M.D., Ph.D.
Pharmakologisches Institut
Universität Heidelberg
Im Neuenheimer Feld 366
6900 Heidelberg/FRG

Dr. DONALD PFAFF, Ph.D.
Rockefeller University
York Avenue, and 66th Street
New York, NY 10021/USA

The picture on the cover has been taken from Nieuwenhuys R., Voogd J., van Huijzen Chr.: The Human Central Nervous System, 2nd Edition. Springer-Verlag Berlin Heidelberg New York 1981

ISBN-13: 978-3-642-69730-2 e-ISBN-13: 978-3-642-69728-9
DOI: 10.1007/978-3-642-69728-9

Library of Congress Cataloging in Publication Data Main entry under title: Actions of progesterone on the brain. Includes bibliographies and index. 1. Brain. 2. Progesterone–Physiological effect. 3. Progesterone–Receptors. I. Ganten, D. (Detlev), 1941– . II. Pfaff, Donald W., 1939– . III. Kato, Junzo. [DNLM: 1. Brain–drug effects. 2. Progesterone–pharmacodynamics. 3. Receptors, Progesterone. Wl CU82Q v. 5 / WP 530 A188] QP376.A24 1985 615'.766 85-8085
ISBN-13: 978-3-642-69730-2 (U.S.)

2121/3130-543210

Contents

Antiprogestins

THOMAS RABE, LUDWIG KIESEL, and BENNO RUNNEBAUM[1]

Contents

Abbreviations
hCG: human chorionic gonadotropin; GnRH: gonadotropin-releasing hormone; LH: luteinizing hormone; ACTH: adrenocorticotropin-releasing hormone

1 Introduction

The most efficient methods of female contraception inhibit ovulation by disturbing the endogenous endocrine-regulating mechanisms. The ovulation inhibitors in current use depend on a constant level of active synthetic hormonal steroids

1 Department of Gynecological Endocrinology, University Hospital for Woman, Vossstraße 9, D-6900 Heidelberg, Federal Republic of Germany

Current Topics in Neuroendocrinology, Vol 5

during the menstrual cycle. This constant hormonal steroid concentration in the serum is achieved either by daily intake or by the use of long-acting depot injections. These methods combine a high efficiency with a relatively low percentage of side-effects. However, they almost always impair the cyclic physiology, thus giving rise to fears of long-term side-effects. Another disadvantage is that an almost continous intake of hormones is necessary, whereby significant quantities of hormones are ingested, although the use of new hormonal stereoids may slightly decrease the side-effects. Safety might be increased if the pharmacological inhibition of the menstrual cycle were reduced to a period of 1 or 2 days.

All the techniques that do not suppress ovulation have a higher failure rate. In recent years there has been a marked increase in the use of mechanical and chemical barrier methods. Since these methods have a higher failure rate there is a need for new contraceptive techniques, and especially of methods that abolish or terminate early pregnancy. These new methods of fertility control have been partially supported by the World Health Organization (WHO). Up to now the methods used for termination of early pregnancy have been uterine aspiration and curettage, which are traumatic both physically and psychologically. Therefore, pharmacological methods are needed. Attempts have been made to intercept early human pregnancy with drugs, such as high-dose estrogens, prostaglandins, vaccines against human chorionic gonadotropins, or GnRH analogs. The success rate of these "interceptives" has remained uncertain, and investigations have been directed at a new approach.

In the menstrual cycle and in the pregnant woman progesterone is a controlling factor. Progesterone is a steroid hormone produced by the corpus luteum during the second phase of the cycle, and in the pregnant woman its formation is increasingly taken over by the placenta. It induces the transformation of the endometrium, which allows the implantation of the embryo. Clinical observations have provided indirect evidence that progesterone is indispensable in woman, as it has been demonstrated in laboratory animals (Csapo and Pulkkinen 1973; Csapo et al. 1974). Thus, it seems progesterone is an ideal target for postconception therapy.

Biologically active progesterone might be reduced by several methods. First, antigonadotropins, antiluteotropins, and luteolytic agents can inhibit progesterone synthesis. One example is provided by oral contraceptives, which, however, require almost continuous intake during the cycle. Second, progesterone synthesis and release can be reduced by enzyme inhibitors of steroidogenesis. These inhibitors are not specific and they affect the production of other important steroid hormones. Third, the transport of progesterone may be impaired by progesterone binders. Clinically, however, it is diffult to imagine the situation with injections of an antibody (passive immunization) to women. Active immunization with antiprogesterone vaccines would affect the entire endocrine system for a long time and cause immunological side-effects. Fourth, the action of progesterone can be blocked at the level of binding sites on end-organs. Progesterone, like other steroids, exerts its action by way of receptors, which bind the molecule and mediate its effect on cell genes. An antihormone that has a similar structure prevents its access to the receptor but fails to activate this binding site and also its action on cellular genes. This mechanism of hormone and antihormone action can be compared with the key and hole theory. While the correct molecule, i.e., the correct key, fits in the hole and can unlock the door, the wrong key fits into the keyhole but cannot unlock the door. The wrong key also prevents the correct key from entering and exerting its action.

This approach has been very promising, and clinical trials have been initiated to test the use of antiprogestins for menstrual regulation and inhibition of implantation.

2 Progesterone

2.1 Progesterone in the Maintenance of Pregnancy

Progesterone is a C_{21} steroid and is secreted in the nonpregnant woman mainly in the corpus luteum of the ovary and, to a lesser extent, in the adrenal. During pregnancy the progesterone synthesis shifts from the corpus luteum graviditatis to the placenta around the 8th or 9th week of gestation (Csapo et al. 1974). Gestationally produced progesterone fulfills two functions. First, the transformation of the endometrium necessary for implantation is induced. Second, progesterone is required during pregnancy to protect the uterus from preterm contractions.

2.1.1 Human Studies

Women experience monthly menstrual bleeding induced by progesterone control. Once pregnancy occurs progesterone is produced continuously by the corpus luteum graviditatis to maintain pregnancy. This has been shown conclusively in animal studies by Csapo and Pulkkinen (1973), and Csapo et al. (1974). These authors demonstrated that luteectomy performed in the first 50 days after conception caused a rapid decrease in serum progesterone and a concurrent increase in uterine contractility leading to abortion. In addition, early pregnancy can be maintained following luteectomy before the 9th week of gestation by the administration of exogenous progesterone (e.g., progesterone suppositories). On the other hand, a sudden local increase in local uterine motility envoked by the intrauterine application of prostaglandins (Csapo 1976) results in suppression of the placental synthesis of progesterone followed by menstrual bleeding (Fig. 1). An explanation of this phenomenon is based on the observation that progesterone synthesis is taken over by the human placenta after the 7th or 8th week of gestation.

2.1.2 Animal Studies

In laboratory animals progesterone is equally necessary for the implantation of embryos and for the maintenance of pregnancy (Fraenkel 1910; Bouin and Ancel 1910; Butenandt et al. 1934).

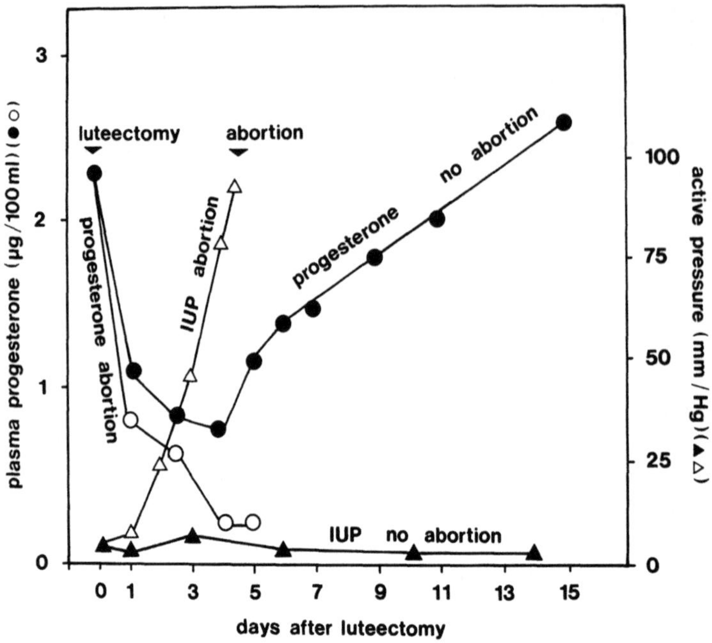

Fig. 1. Influence of luteectomy during early pregnancy on serum progesterone concentrations, intrauterine pressure, and maintenance of pregnancy. (Csapo and Pulkkinen 1978)

The fact that progesterone is required for the maintenance of pregnancy implies that any mechanism that interferes with the synthesis or the action of progesterone in early gestation can cause abortion. There are several substances that can provoke such an interference: first, compounds that inhibit progesterone synthesis in either the corpus luteum or the human placenta; second, substances that counteract the action of progesterone at the myometrial receptor sites; and third, a group of compounds that cause direct contractions of the uterus. The first group inhibits enzyme systems of the progesterone synthesis. The second group is made up of antigestagens and the third, of prostaglandins acting directly on the uterine muscle cells.

Substances used for the induction of abortion in early pregnancy vary in their site of action according to gestational age. Before the 8th or 9th week of gestation compounds stop progesterone synthesis in the corpus luteum by luteolysis. After the 9th week of gestation steroidogenic blockers of placental hormone production are required.

Ovarian cells start to secrete progesterone around the ovulatory period. Different schemes have been observed according to the animal species, but some progesterone is always produced and released into the blood before endometrial cells are transformed for implantation, indicating that progesterone may play an important role in this process (Short 1972).

3 Progesterone Receptors

3.1 Receptor Theory

Until recently the major classes of steroid hormones (estrogens, gestagens, androgens, glucocorticoids, and mineralocorticoids) were thought to act according to the following model:

Steroid hormones bind to the specific cytoplasmic proteins or receptors, followed by activation of the complex and translocation to the nucleus. The complex binds to the nuclear receptor sites on target cell chromatin, thereby modulating the activity of specific genes responsible for the synthesis of particular species of mRNA. The cytoplasmic steroid hormone-receptor complexes undergo a hormone-induced conversion, termed receptor activation, which causes the accumulation and binding in the nucleus.

The interaction of steroid hormones with cellular components depends on the amount of free hormone available to the cell. In the blood, steroids are bound to specific and nonspecific components. These steroid-binding plasma proteins may also be present in uterine extracts (Milgrom and Baulieu 1970b). Steroid hormones enter cells by diffusion. They bind within target cells to soluble macromolecules called receptors. These receptors are characterized by high affinity for the active hormone(s) and not for other steroids, and they are believed to mediate most if not all hormone action at the cellular level.

However, recent findings (Gase et al. 1984) in the chick oviduct have demonstrated that the concept of steroid action that necessarily involves formation in the cytoplasm of hormone complexes may have to be revised. The authors suggest that the necessary and sufficiently active hormone-receptor complexes may be formed in the nucleus in which the unoccupied receptor is originally present and where gene response will take place.

By definition, a hormone receptor has two "sites" one is designated "receptive" (r), and is the site at which the hormone binds, while the other is called the "executive" (e) site and initiates the hormonal effects (Baulieu et al. 1971) (Fig. 2).

Progesterone and progesterone-like compounds bind specifically to the receptor's receptive site, for which they have a high affinity (Milgrom and Baulieu 1970a; Wiest and Rao 1971): there is stereospecificity and biospecificity. When the hormone has bound the receptive site r the receptor undergoes a conformational change allowing an interaction of the executive site e with some "acceptor" component of the cell, presumably at the chromatin level, which in turn triggers a metabolic modification. The target cell acceptor reached and modified by the progesterone-receptor complex is still unknown in molecular terms.

Besides the two sites r and e, the fact that progesterone leads to receptor inactivation can be described phenomenologically in terms of an "inactivation site." The latter, operating secondarily to progesterone arrival in target cells, may correspond to a variety of mechanisms, leading to reversible or irreversible inactivation (for instance by proteolysis). Irreversible inactivation could result from a specific conformational change due to progesterone (r on Fig. 2) making part

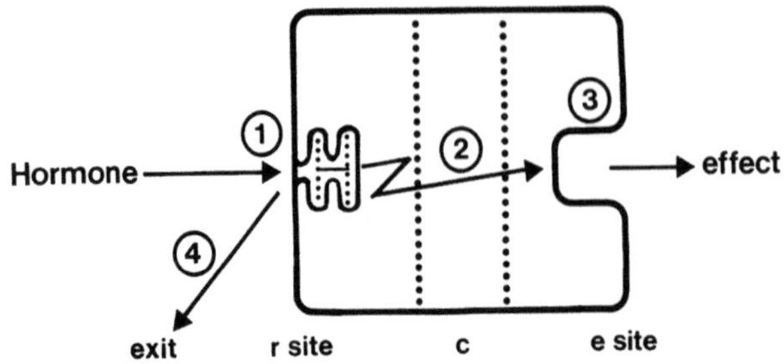

Fig. 2. A steroid hormone receptor. *Hormone* means any steroid hormone. It binds to the receptive (r) site (*1*) with high affinity and transduction takes place (*2*) coupling the hormone binding to activation of the executive (e) site (*3*). The e site may interact with another macromolecule and catalyze some reaction. When this effect is initiated the hormone cannot do anything but leave the receptor (*4*). (Milgrom et al. 1972b)

of the polypeptide chain of the receptor susceptible to a protease. Alternatively, progesterone could directly (c) or indirectly (e) activate a proteolytic enzyme specific for the progesterone receptor. Reversible inactivation might result, for example, from interaction with the receptor of a product of the metabolism as modified by progesterone.

Finally, then, operationally a schematic (Fig. 2) can be justified from which it is clear that binding of progesterone to site r leads to activation of sites e, showing a sort of built-in and balanced relation between events initiated at site e (effects) and inactivation, as if the physiological hormonal ligand binding were simultaneously providing an effect by transduction and a negative control by way of receptor inactivation.

3.2 Physiology

Some features of the hormonal control of the cyclic progesterone receptor content of the uterus have been studied (Milgrom et al. 1973a). The increase during proestrus is probably attributable to estrogens, since in castrated (Milgrom et al. 1970) and prepubertal (J. P. Raynaud and D. Philibert, personal communication) animals estradiol provokes a significant augmentation of binding sites. Such an increase in the density of receptors may be part of the molecular mechanism of the classic potentiation of progesterone action by previously administered estradiol (Courrier 1945). The estradiol induction of progesterone receptor can be suppressed by protein and RNA synthesis inhibitors (Fig. 3). The apparent decay of the progesterone receptor induced by estradiol in castrated guinea pigs corresponds to a half-life of at least 5 days and is not dependent on protein synthesis. In this model situation, where hormonal manipulations are performed easily, progesterone can be injected when the level of receptors present is at its maximum: in this case, the decay of the progesterone receptor binding sites is accelerated, only 20% remaining measurable after 1 day. This faster decrease does not reflect masking or nuclear translocation of the receptor, nor is it dependent on protein synthesis. From these observations it follos that during the guinea pig estrus cycle,

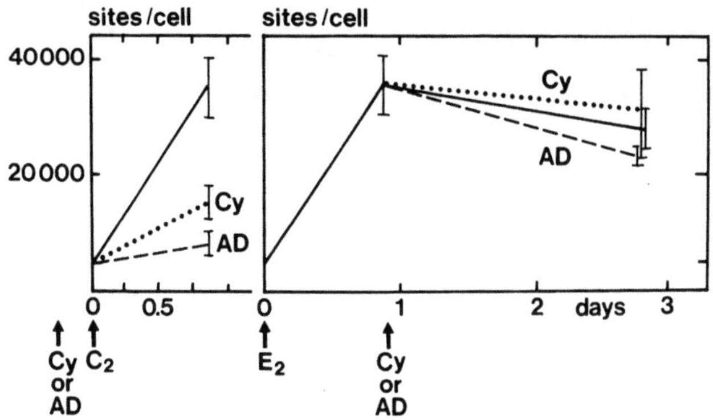

Fig. 3. Induction of progesterone receptor by estrogen in castrated guinea pig uterus. The maximum level is at about 1 day after injection of estradiol (E_2). The negative effects of protein and RNA synthesis inhibitors (*Cy* and *AD*) are shown. The prolonged apparent disappearance of the receptor (half-life 5 days) is shown in the *right-hand panel* and would not account for the physiological decrease in receptor observed between days 0 and 4 of the cycle. (Milgrom et al. 1973 a)

it is logically possible to attribute the rapid decrease of the progesterone receptors after proestrus to the progesterone of the first and possibly of the early part of the second luteal peaks. In the diestrus period the progesterone receptor level is even lower than that measured in castrated animals not treated with hormones (Milgrom et al. 1972 a), which might explain some "negative" effects of progesterone. Up to no now it has been impossible to decide whether the low level of progesterone receptors in the uterus during the luteal phase is due only to the effect of periovulatory progesterone, inactivating the receptors induced by estrogen at proestrus, or whether abundant synthesis of receptors takes place continuously and these are rapidly destroyed by the progesterone secreted during the luteal phase. Experiments with agents neutralizing specifically circulating estradiol and progesterone, e.g., antibodies, might provide some information on this point.

A series of recent studies in the guinea pig suggests that the progesterone receptors of the uterus (Milgrom et al. 1970) depend on some complex hormonal control during the estrus cycle (Milgrom et al. 1972 b). In Fig. 4 the progesterone levels in the plasma (Feder et al. 1968) and the progesterone receptor concentration per uterine cell are plotted (Milgrom et al. 1972 b); the values for the latter refer to the determination of high-affinity specific binding sites for progesterone of both endometrium and myometrium, whether they are free (amenable to direct labeling with radioactive hormone) or occupied by the endogenous hormone secreted during the estrus cycle. [the method for measuring occupied binding sites is based on simple physical chemical principles, reported by Milgrom et al. (1972 a)]. Plasma progesterone and uterus progesterone receptors show cyclic variations, which do not coincide. Progesterone exhibits a first increase at ovulation time and a prolonged high plasma level during diestrus (luteal phase). On the other hand, there is a peak of receptor, rapidly developed at the proestrus phase. The maximum value is not maintained for very long and a decrease follows, so that the receptor level is very low during the luteal phase (diestrus) even though

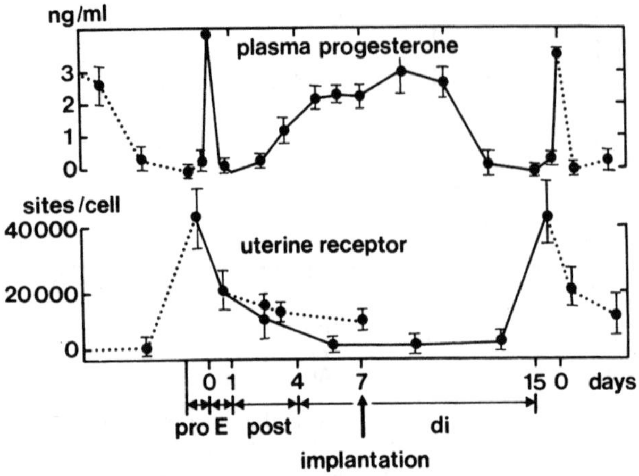

Fig. 4. Changes in, and hormone control of, progesterone receptor in guinea pig uterus, showing plasma levels of progesterone over the cycle and concentration per cell of the uterine progesterone receptor. *pro*, proestrus; *E*, estrus; *post*, after estrus. The diagram illustrates the situation during pregnancy. (Milgrom et al. 1972b)

this is also the period at which implantation eventually takes place. Incidentally, in the case of pregnancy the receptor concentration is similar to that found in the absence of fertilization up to implantation (Fig. 4).

These observations suggest that if the progesterone receptors are implicated in processes leading to implantation it may be the progesterone available when its receptors are high, around ovulation, that is physiologically important. There is actually circumstantial evidence for this, since Deanesly (1960) obtains some successful implantations in the guinea pig even after ovariectomy, provided the latter is performed more than 3 days after ovulation (implantation takes place on the 7th day after ovulation).

These results are the experimental background to the proposed method for fertility control. In the hamster, recent results are similar to those obtained in the guinea pig (Leavitt et al. 1974). In women, preliminary results indicate, as expected, that there is progesterone-binding protein in the uterus (Wiest and Rao 1971), which is increased by estrogen treatment. The part played by transcortin in progesterone binding in extracts from human material is quantitatively still undecided. In the endometrium, progesterone binding increases in the late proliferative phase of the menstrual cycle (Haukkamaa et al. 1971). Another indication comes from the observation of a higher concentration of progesterone in the endometrium at midcycle: this observation is also compatible with (but does not prove) a periovulatory increase of receptor-binding sites, especially if one recalls that there is a preovulatory increase but no clear periovulatory peak of plasma progesterone in women. At most, there is an initial burst on the ascending slope of progesterone plasma concentration starting after LH discharge (Ross et al. 1970; Van De Wiele et al. 1970).

3.3 Pharmacology

3.3.1 Binding and Activity

In many species (mouse, rat, guinea pig, hamster, sheep, calf, and human), high-affinity receptors for progesterone have been identified in the uterine cytosol (Milgrom and Baulieu 1970a; McGuire and De Della 1971; Thomas 1973; Leavitt et al. 1974; Guériguian et al. 1974; Kontula 1975b). The complex steroid and receptor is heat and acid labile (Jänne et al. 1975), and the binding activity of progesterone to its receptor can be inhibited by sulfhydryl-blocking agents, indicating that SH groups are involved in the formation of the steroid-receptor complex (Kontula et al. 1974a; Jänne et al. 1975). The association constant for progesterone to the progesterone receptor is approximately 10^9 to 10^{10} liter/mol. A sedimentation coefficient of 6–7 S or 4–5 S has been determined in density gradient buffer. The 6–7 S receptor is characterized by a highly specific affinity for progesterone (Kontula et al. 1973, 1974b). However, some species differences have been revealed (Kontula et al. 1974b; Kontula et al. 1975). In addition, the ligand specificity of progesterone receptors isolated from different tissues (cervix and vagina) has been found to be very similar to that of the uterine receptor (Atger et al. 1974), and no difference was been found between myometrial and endometrial receptor (Jänne et al. 1975; Luu Thi et al. 1975; Kontula 1975b).

The development of antiprogestins has attracted a great deal of interest with a view to the development of new contraceptive compounds. Initially the alkylation of receptor proteins was attempted (Clark et al. 1974, 1975). However, alkylation on the receptor did not occur with diazoketone derivatives of 17-hydroxyprogesterone. Beyer et al. (1976) succeeded in synthesizing alkylated derivatives of 17-hydroxyprogesterone, which had a low affinity for progesterone receptor, making it unlikely that they would be active as antiprogestin compounds. Receptors closely resembling the uterine binding component have been identified in the cervix and vagina (Atger et al. 1974). Similarly, the myometrial and endometrial progesterone receptors are thought to be very similar in such species as guinea pig, human, and sheep (Jänne et al. 1975; Luu Thi et al. 1975; Kontula 1975b).

3.3.2 Number of Receptors

The synthesis of the 6–7 S receptor is under hormonal control. The uterine concentration of this receptor declines dramatically following ovariectomy, but rapidly increases again in response to subsequent estrogen treatment (Thomas 1973; Jänne et al. 1975; Milgrom et al. 1973b). In contrast to the stimulative effect of estrogen, progesterone induces a marked decline in the number of uterine receptor sites (Milgrom et al. 1973b). Determination of uterine 6–7 S receptor concentration during the estrous cycle of the hamster and guinea pig has revealed characteristic changes which reflect the cyclic fluctuations in ovarian steroid production (Leavitt et al. 1974; Milgrom et al. 1973b). Hence, receptor concentration is at its maximum during proestrus, when plasma estrogen levels are high; and at its minimum during diestrus, when estrogen levels are low and progesterone is being secreted by the corpora lutea. Similarly, the concentration of binding sites in human myometrial and endometrial tissue reaches a maximum during the late proliferative phase as a consequence of the preovulatory rise in ovarian estrogen secretion (Bayard et al. 1975; Kontula 1975c). Estrogen administration to postmenopausal women also induces the appearance of a 6–7 S progesterone-binding receptor in myometrial cytosol (Jänne et al. 1975).

Both RNA and protein synthesis appear to be involved in the estrogen-induced increase in 6–7 S receptor concentration (Milgrom et al. 1972a, 1973a). The suppressive effect of progesterone treatment is associated with the disappearance of the 6–7 S binding component from the uterine cytosol, and may be the result of an inhibition of uterine RNA synthesis (Trams et al. 1973) or the destruction of the receptor with proteolytic enzymes.

In addition to the 6–7 S binding component, there is some evidence for another category of receptor sedimenting at about 4 S. In the normal cyclic human uterus, for example, progesterone receptors have been identified in both the endometrium and myometrium with sedimentation coefficients of 3.7–3.8 S (Kontula 1975c; Smith et al. 1975). The pregnant guinea pig uterus also possesses a 5 S progesterone receptor, which contrasts with the 7 S binding component induced by estrogen (Kontula et al. 1973). It has been suggested that the estrogen-induced 6–7 S component represents a complex of the true receptor and another estrogen-dependent macromolecule (Kontula 1975a).

4 Test Models for Antigestagens

4.1 Preconditions

The biological evaluation of new compounds includes the identification of characteristic antiprogestational actions as well as the determination of specificity and mechanism of action. In general, the antigestagenic activity of a substance is characterized in vitro by means of receptor studies (competitive displacement from the receptor), and by numerous biological test models in vivo. A precondition of such experiments with antiprogestins in animal models is the progesterone dependency of the respective biological response: the maintenance of pregnancy in all experimental animals, the support of deciduation of rodent uterus, and endometrial proliferation in rabbits. These responses can be provoked by exogenous as well as endogenous progesterone. Studies with active antiprogestins in animals in comparison with exogenous and endogenous progesterone give a good indication of the mechanism of action. Antigestagenic compounds can be separated into those inhibiting the production of progesterone and others that inhibit its action. The first inhibit the action of exogenous progesterone, but they are ineffective if exogenous progesterone is given. On the other hand, antiprogestins, which counteract progesterone, inhibit the action of both exogenous and endogenous progesterone.

In conclusion, studies that aim to clarify antiprogestin actitivity of either type should be initiated with endogenous progesterone. If distinction of the mecha-

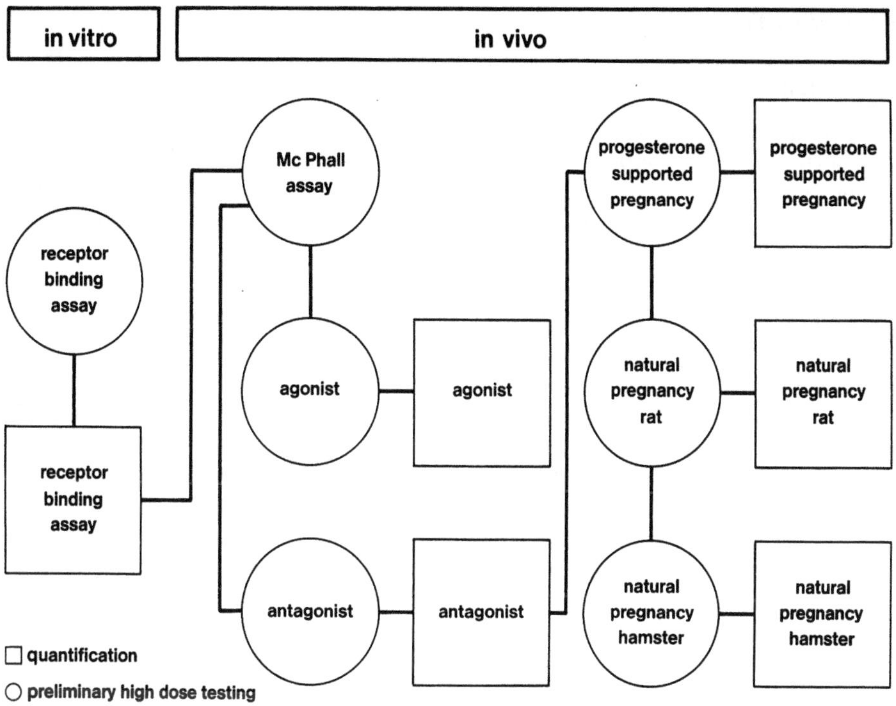

Fig. 5. Preclinical development scheme for competitive progesterone antagonists with abortifacient activity. The assays shown in *circles* are performed at a single high concentration (in vitro) or dose (in vivo) of the compound, whereas those shown in *squares* are performed at multiple concentrations or doses to quantify the results. (Reel et al. 1979)

nism of action is required, exogenous progesterone models should be used to determine whether the compound inhibits the action of progesterone or whether it merely inhibits progesterone synthesis.

4.2 In Vitro Tests

Steroids with antihormonal activity have been found mainly in mineralocorticoids and estrogens.

On the basis of the concept of the mechanism of steroidal antihormone action, it can be assumed that active antiprogestins will have an affinity for progesterone receptor but that the biological response will be inhibited with little or low gestagenic effect.

In studies with humans, sheep, and rabbits a reasonable correlation between binding affinity and uterine receptor protein has been demonstrated for gestagen agonists (Kontula et al. 1975). Apparently various characteristics of the steroids are necessary, such as a pregnane or estrane steroid skeleton or a Δ^4-3-keto group, and major bonds between steroid and receptors are hydrophobic (Luu et al. 1975). In summary, the basic structure of an active antagonist compounds seems to resemble that of the agonist.

4.3 In Vivo Tests

4.3.1 Animal

In most rodents and Lagomorpha sensitivity towards estrogens and continuous administration of low-dose estrogen are necessary for an optimum response to progestins. It is well known that estrogens not only potentiate the response to progestins but may also diminish its effect. High doses of estrogens have an antiprogestin effect in rabbits (Chambon 1949) and they also cause inhibition when given to progesterone-treated ovariectomized rats, by way of a peripheral antagonism (Neumann and Elger 1971). Receptor studies have explained some aspects of these observations. The number of progesterone-binding sites in the rat uterus is increased by the treatment with estradiol (Rao et al. 1973).

In contrast to estrogens, an optimal response to progestins can be achieved in rodents and Lagomorpha by a surplus of progesterone. Relatively small doses of estradiol are sufficient to give optimum conditions for the action of progesterone, whereas high estradiol doses inhibit progestin response. In certain circumstances such an inhibition can be overcome by the increased progesterone.

Estrogen stimulation and the renewal of progesterone receptors have not been sufficiently clarified for a full analysis of estrogen–progestin interaction. In primates there is only limited knowledge about estradiol–progesterone interactions in the endometrium. In comparison with the rat the endometrium of the primate reacts more extensively to progesterone and more intensely to estradiol. In rhesus monkeys, for example, similar doses of progesterone per animal but much higher doses of estrogens are required to achieve an optimum response of the decidua (Good and Moyer 1968).

However, further experiments on the activity of potential antigestagens are necessary in primates to validate the routine experimental approach in rabbits and rats. Since testing of compounds from pregnant monkeys is mainly a logistic problem, it would be of interest to test the activity of compounds in monkeys during pseudopregnancy induced by continuous administration of HCG in the luteal phase.

It has been known for some time (Brown and Bradbury 1947; Fried and Rakoff 1952) that women experience pseudopregnancy if HCG is administered continuously after the 20th day of the cycle to prolong the action of the corpus luteum. Daily doses of 5,000 IU HCG or more are required to prolong the action of the corpus luteum over 13 days; this can be monitored by recording the excretion of pregnandiol and the length of cycle.

Treatment with HCG has been applied to test the abortifacient action of steroids that interfere with the function of the human corpus luteum (e.g., luteolytic drugs). The effect of oxymetholone, for example, a luteolytic agent with no effect in early pregnancy, is abolished if HCG is administered in the prolonged luteal phase (Henzl et al. 1973). This study indicates the necessity of using the HCG test to monitor compounds for postconception fertility control.

4.3.2 Human

In primates the antigestagen effect of higher estrogen doses given together with progesterone during the proliferative phase has not been fully analyzed. However, it is known that extremely high doses of estrogens (ethinylestradiol up to 5 mg/day) do not terminate early pregnancy in women (Bacic et al. 1973).

Estrogens, which have potent antiprogestin properties in the rabbit and rodents and do not induce menstrual bleeding, have been studied with reference to estrogenic effect. The antiprogestin effect of estrogens has been simulated in all experimental models used.

4.3.2.1 Midcycle Contraception

It can be assumed that there is some obligatory linkage between progesterone availability, progesterone binding, progesterone effects, and inactivation of the progesterone receptor, and that a decline in receptor density changes the "receptivity" of uterine cells to further progesterone exposure. Therefore, since concentrations of progesterone undergo cyclic changes, the level of available circulating progesterone during the estrus cycle may be crucial to the preparation of uterine cells for successful implantation. It follows that the introduction of an excess of progesterone at midcycle, for example by days 15–16, when the amount of progesterone secreted is increasing but still rather low, may lead to a premature effect, including inactivation of the receptor; one consequence of this too early progesterone would be an abnormal "secretory" transformation phase of the endometrium.

All derivatives that have any progesterone-like activity "must" bind to the receptive site of the receptor. Indeed, several reports have indicated that this is the case (Milgrom et al. 1970a; Wiest and Rao 1971). However, since these compounds are structurally somewhat different from progesterone, it is conceivable that the consequences of their binding at the r site could be different from those of progesterone binding. A change in favor of receptor inactivation (following the three-site theory) or even a relatively low activation of the e site (ordinary two-site model of receptor) might be responsible for an antiprogesterone effect. This could be the case with the synthetic derivative R-2323, which displays intrinsic antiprogesterone properties (Sakiz and Azadian-Boulanger 1971) and which, when given to women at days 15–17 of the cycle in competition with endogenous and physiologically significant progesterone, could prevent its action and consequently also prevent successful implantation (Sakiz et al. 1974).

With any method of administration including the use of orally active preparations or of vaginal rings, a major difficulty may be encountered as a result of the physiological irregularity of the human menstrual cycle. Therefore, clinical studies with the compound given when estrogen-induced receptor synthesis is complete might be conducted with the help of the available methods of measuring estrogen or LH, cervical mucus changes, or even temperature rise.

Whether or not they lead to a midcycle contraceptive technique, the above mechanisms may be partly responsible for the contraceptive effects of other methods based on progesterone, progestagens, or antiprogesterone derivatives. They could be involved when low doses are given continuously, either systemically or locally with medicated intrauterine devices. They may also be operative in postcoital methods, when the compound is given occasionally (Johansson 1971).

4.3.2.2 Luteolysis and Early Pregnancy Wastage

The normal survival period of the corpus luteum without pregnancy is generally about 14 days. For reasons that are not yet known, a decrease in progesterone secretion and a resulting progesterone withdrawal bleeding (menstrual bleeding) occur thereafter. Progesterone secretion is stimulated by the pregnancy-specific hormone HCG, which is secreted during tubal passage and prolongs the last phase of the corpus luteum. This extension, however, is not unlimited, and after the 7th or 8th week of gestation a physiological luteolysis of the corpus luteum graviditatis occurs. At this time the placenta supplies progesterone in increasing amounts; these are required for the maintenance of pregnancy.

The implantation of the human blastocyst as well as the process of early pregnancy before the 7th to 8th week of gestation can be interfered with in three main ways (Aitken and Harper 1977). First, luteal depression can be achieved by means of prostaglandins or steroids. Second, inactive progestins can occupy the progesterone receptors on the endometrium and the myometrium and inhibit the effect of gestagens. Third, the inhibition of early luteotrophic activity of the blastocyst, i.e., HCG secretion, results in the indirect inhibition of progesterone production in the corpus luteum.

In primates the early gestational product has a direct luteotrophic, and possibly an antiluteolytic, effect on the corpus luteum (Knobil 1973). It seems likely that the luteotrophic factor is HCG, which increases in the serum of pregnant women after about the day after implantation (Jaffe et al. 1969; Saxena et al. 1974). HCG causes a rapid increase of plasma progesterone levels (Hanson et al. 1971; Niswender et al. 1972). Similar increases in plasma progesterone (Neill et al. 1969) have been observed in rhesus monkeys following implantation (Reinius et al. 1973) and following the appearance of choriogonadotropin (Meyer 1972). In some other species, such as rabbit (Fuchs and Beling 1974), rat (Haour 1976), and mouse (Beyer and Zeilmarker 1974), the fetoplacental unit produces gonadotropic hormones.

A uterine antiluteolytic substance (Anderson et al. 1972; Short 1969) produced by the blastocyst has been demonstrated in guinea pig, hamster, pig, rabbit, rat, and sheep. Its role as an antiluteolytic substance is not yet known.

In principle, any substance that is produced in the blastocyst and exerts a vital function on the implantation and support of the corpus luteum is potentially a site of action for contraceptive substances. If its action is blocked pregnancy can be prevented at the implantation stage without affecting the menstrual cycle. The β-HCG antibodies are one group of such substances.

A further approach in fertility control is the development of antihormones, which compete with HCG for receptor sites of the corpus luteum. Such compounds could be used if menstruation fails to occur, or possibly at the time in the cycle at which menstruation is expected, to guarantee regression of the corpus luteum. Yang et al. (1976) have found an LH receptor-binding inhibitor (LHRBI) in aqueous extracts in luteinized rat ovaries. A similar LH-HCG receptor-binding inhibitor has been identified in pig corpus luteum extracts (Sakai et al. 1977).

Finally, a competitive inhibitor of HCG has been developed by enzymatic cleavage of carbohydrate moieties, which lacks the luteotrophic activity of the native molecule (Bahl 1969). It has been shown that with the removal of sugars and sialic acids (galactose, *n*-acetyl-glucosamine, and mannose) the receptor affinity of HCG in vitro is still affected, whereas the ability to stimulate cAMP production in granulosa cells is abolished (Channing et al. 1976). The activity of such derivatives as antigonadotropins has been demonstrated by inhibition of HCG-induced progesterone secretion in pig and monkey granulosa cells in vitro (Channing et al. 1977, 1978). Further questions regarding immunogeneity of such molecules and the biological half-life are currently being researched.

Estrogens can cause luteolysis in a dose-dependent manner in laboratory animals (guinea pig, hamster, rabbit, rat) and in several domestic animals (cow, sheep, pig) (Anderson et al. 1972; Oriol-Bosch and Cortes 1975).

High doses of estrogens have a potent antifertility effect in women. There are indications that estrogens given at a dose of 5 mg over 5 days postcoitally at mid-cycle inhibit implantation and thereby pregnancy by way of a luteolytic mechanism of action. Proving this point, estrogens were found to cause a dose-dependent decrease of plasma progesterone levels and a shortening of the cycle when administered in the luteal phase. The luteal depression was also induced by the implantation of estrogen crystals into the corpus luteum. No signs of luteal depression were found, however, following estradiol benzoate and diethylstilbestrol treatment (Oriol-Bosch and Cortes 1975). It must be borne in mind that the dose and/or potency of the estrogens that were used in these experiments were lower than those in the studies where luteolytic and contraceptive effects were observed.

Theoretically, any compound not leading to effects while competing with progesterone for the r site of the receptor (Fig. 2) is a progesterone antagonist and potentially usable at the end of the luteal phase in women as a "period inducer," whether or not fertilization has taken place. The amount of endogenous hormone to be counteracted is relatively great (in the case of pregnancy especially) and implantation may be well established, so that massive administration of an antiprogesterone compound for several days would be needed to counteract the progesterone maintaining the pregnancy. The possibility of devising an affinity-labeling antiprogestin has been considered (Warren 1973).

The possible antiprogesterone effects of progesterone and progesterone analogs are not the only way of abolishing progesterone action by way of an effect on the progesterone receptor. For instance, an antiestrogen given during the first part of the estrus cycle could lead, among other things to a decreased level of progesterone receptor induction. On the other hand, estrogens themselves are antiprogesterone compounds when given simultaneously with progesterone (review in Courrier 1945). However, it is probable that the mechanism(s) do(es) not involve receptor control.

Finally, the possibilities of drastically reducing progesterone secretion (i.e., using GnRH antagonist, estrogen, or progestagen during the luteal phase, or enzyme inhibitors, luteolytics, prostaglandin, etc.) obviously fall in another class of methods, which aim at changing the hormone supply and not the receptors.

5 Antiprogestins

5.1 Structures

Alkylated derivatives of 17-acetoprogesterone were used to test the hypothesis that bulky groups in certain positions of the steroid molecule can convert progestins to antiprogestins. These groups change the molecule binding outside the area of the actual progesterone-binding site. These compounds were tested against radioactive tritiated progesterone in view of their competitive affinity for cytosolic receptors of rabbit uterus. It seems unlikely, however, that these compounds can act as antiprogestins, because of their low affinity.

5.2 Pharmacological Action In Vitro

A specific antiprogestin has been developed by Baulieu's group and the Roussel-Uclaf Company. This progestin blocks the action of progesterone, which is formed in the corpus luteum or the placenta, and of the receptors in the endometrium and myometrium. According to the pharmaceutical company no systemic side-effects have been observed.

Progestagens

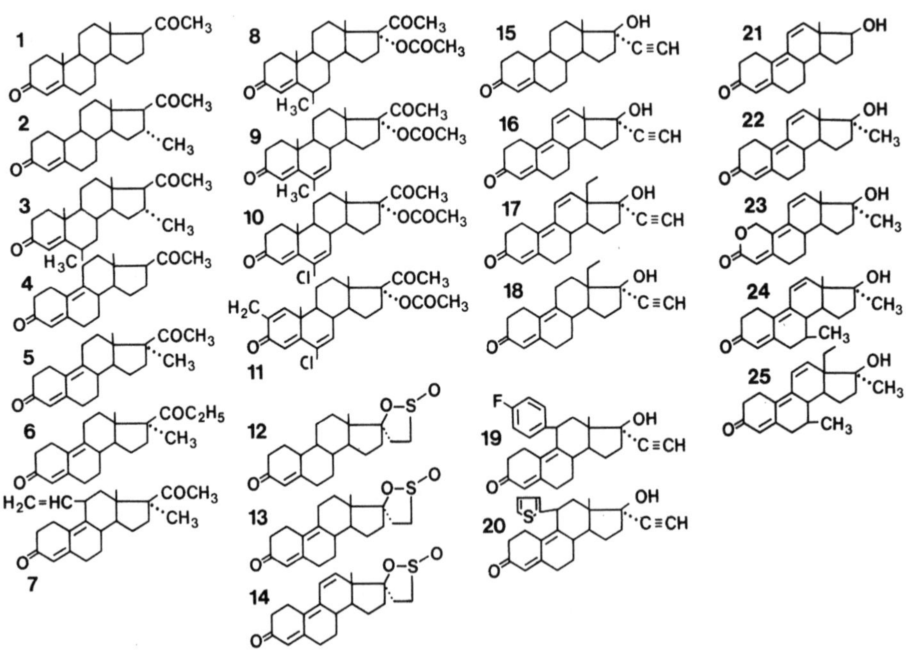

Fig. 6. Structures of synthetic progestagens with agonist and antagonist activity. The chemical names of the individual steroids are given in Table 1

Table 1. Chemical names of test steroids shown in Fig. 6

1. Progesterone	Pregn-4-ene-3,20-dione
2. 16α-Methylprogesterone	16α-Methyl-pregn-4-ene-3,20-dione
3. 6α,16α-Dimethylprogesterone	6α,16α-Dimethyl-pregn-4-ene-3,20-dione
4. RU 3,163	19-Nor-pregna-4,9-diene-3,20-dione
5. Demegestrone (RU 2,453)	17α-Methyl-19-nor-pregna-4,9-diene-3,20-dione
6. Promegestone (RU 5,020)	17α,21-Dimethyl-19-nor-pregna-4,9-diene-3,20-dione
7. RU 25,253	11β-Ethinyl-17α-methyl-19-nor-pregna-4,9-diene-3,20-dione
8. Medroxyprogesterone acetate	17α-Acetyloxy-6α-methyl-pregn-4-ene-3,20-dione
9. Megestrol acetate	17α-Acetyloxy-6-methyl-pregna-4,6-diene-3,20-dione
10. Chlormadinone acetate	17α-acetyloxy-6-chloro-pregna-4,6-diene-3,20-dione
11. Cyproterone acetate	17α-acetyloxy-6-chloro-1,2-dihydro-(1β,2β)-3'H-Cyclopropa(1,1)pregna-1,4,6-triene-3,20-dione
12. RU 22,779	(17R)2'-Oxydo-spiro(oestr-4-ene-17,5'(1,2)-oxathiolan)3-one
13. RU 23,747	(17R)2'-Oxydo-spiro(oestra-4,9-diene-17,5'(1,2)-oxathiolan)3-one
14. RU 25,051	(17R)2'-oxydo-spiro(oestra-4,9,11-triene-17,5'(1,2)-oxathiolan)3-one
15. Norethisterone	17β-Hydroxy-19-nor-17α-pregn-4-en-20-yn-3-one
16. Norgestrienone	17β-Hydroxy-19-nor-17α-pregna-4,9,11-trien-20-yn-3-one
17. Norgestrel	13β-Ethyl-17β-hydroxy-18,19-dinor-17α-pregn-4-en-20-yn-3-one
18. Gestrinone (RU 2,323)	13-Ethyl-17β-hydroxy-18,19-dinor-pregna-4,9,11-trien-20-yn-3-one
19. RU 25,593	11β-(4-Fluorphenyl)-17β-hydroxy-19-nor-17α-pregna-4,9-dien-20-yn-3-one
20. RU 25,055	17β-Hydroxy-11β-(2-thienyl)-19-nor-17α-pregna-4,9-dien-20-yn-3-one
21. Trenbolone (RU 2,341)	17β-Hydroxy-oestra-4,9,11-trien-3-one
22. Metribolone (RU 1,881)	17β-Hydroxy-17α-methyl-estra-4,9,11-triene-3-one
23. RU 2,999	17β-Hydroxy-17α-methyl-2-oxy-oestra-4,9,11-trien-3-one
24. RU 2,420	7α,17α-Dimethyl-17β-hydroxy-oestra-4,9,11-trien-3-one
25. RU 4,841	7α,17α-Dimethyl-13β-ethyl-17β-hydroxy-gona-4,9,11-trien-3-one

The in vitro potency of antiprogestins can be investigated by means of the competitive receptor binding to a progesterone receptor (Table 2) and of the ACTH release in rat anterior pituitary cells in culture (Table 3).

5.3 Pharmacological Action In Vivo

5.3.1 R 2323

R 2323 behaves as an antiprogestin in relation to exogenous and endogenous progesterone (Sakiz and Azadian-Boulanger 1971).

Table 2. Receptor binding of antiprogestin to progesterone receptor (0 °C) after different incubation times (Raynaud et al., 1981)

	2 h	24 h
(1) Progesterone	100	100
(2) 16α-Methylprogesterone	65	60
(3) 6α,16α-Dimethylprogesterone	50	155
(4) RU 25,253	200	245
(5) Demegestone	230	420
(6) Promegestone	220	535
(7) RU 25,253	180	530
(8) Medroxyprogesterone acetate	125	305
(9) Megestrol acetate	150	120
(10) Chlormadinone acetate	175	320
(11) Cyproterone acetate	80	60
(12) RU 22,779	205	335
(13) RU 23,747	225	380
(14) RU 25,0051	330	870
(15) Norethisterone	155	265
(16) Norgestrienone	65	45
(17) Norgestrel	170	905
(18) Gestrinone	75	50
(19) RU 25,593	40	35
(20) RU 25,055	70	85
(21) Trenbolone	75	15
(22) Metribolone	210	190
(23) RU 2,999	260	305
(24) RU 2,420	280	330
(25) RU 4,841	230	675

Animal Experiments. This compound was used to demonstrate antiproliferative activity in rabbits and antidecidual activity in rats following treatment with progesterone. The determination of pregnancy in normal and ovariectomized rats, after treatment with exogenous progesterone, was achieved by R 2323. Ovulation was suppressed in mature rats with normal cycles, but not in premature gonadotropin-treated rats. The transport of the ovum remained unaltered. The proliferative response in rat uterus was counteracted by the administration of R 2323 or antiestrogens during the estrogen peroid. R 2323 remained ineffective when used together with progesterone. In studies with cytosolic receptor proteins estradiol did not compete with progesterone for binding to the estrogen receptor and no compound competed with estradiol for binding to the estradiol receptor; however, there was competition for the progesterone receptor, in the absence of any correlation with progestin or antiprogestin activity.

In animal experiments with R 2323 a marked antiprogestin activity was found; this was independent of the weak antiestrogen activity but might have been related to the weak progesterone activity (Azadian-Boulanger et al. 1976). Receptor binding studies have shown that R 2323 competes with progesterone

Table 3. Agonist and/or antagonist activity of "progestins" on ACTH release in rat anterior pituitary cells in culture (Raynaud et al., 1981)

	Agonist activity ED_{50} (n*M*)	Antagonist activity K_D (n*M*)
(1) Progesterone	300	1,000
(2) 16α-Methylprogesterone	0	200
(3) 6α,16α-Dimethylprogesterone	30	0
(4) RU 3,163	300	300
(5) Demegestone	1,000	300
(6) Promegestone	0	400
(7) RU 25,253	30	300
(8) Medroxyprogesterone acetate	100	300
(9) Megestrol acetate	900	400
(10) Chlormadinone acetate	1,000	200
(11) Cyproterone acetate	0	300
(12) RU 22,779	0	250
(13) RU 23,747	0	150
(14) RU 25,051	10,000	1,000
(15) Norethisterone	100,000	0
(16) Norgestrienone	1,000	0
(17) Norgestrel	0	300
(18) Gestrinone	0	300
(19) RU 25,593	ND	ND
(20) RU 25,055	ND	ND
(21) Trenbolone	1,000	100
(22) Metribolone	0	1,000
(23) RU 2,999	0	300
(24) RU 2,420	300	300
(25) RU 4,841	500	400

and with the uterine cytosolic protein, but not with estradiol (Sakiz et al. 1972; Philibert and Raynaud 1974). These observations are the basis for the assumption that R 2323 given after ovulation saturates the progesterone-binding sites at the beginning of the luteal phase and causes progesterone-dependent changes, thereby inhibiting implantation.

Further studies on the cytosolic receptor (Raynaud et al. 1975) expanded and improved the methods of characterizing the interactions between steroids and cytoplasmic receptors. Tissues have been used to determine the binding of estrogens to the mouse uterus, androgens to the prostate, progesterone to the rabbit uterus, and mineralocorticoids to the rat kidney. It was observed that compounds with moderate affinity for receptors are agonists, whereas those with weak affinity are weak agonists with an antagonistic effect. R 2323 did not compete with progesterone for the progesterone receptor but did compete for the androgen and aldosterone receptors.

Clinical Trials. In the human, R 2323 administered according to various regimens was shown to be effective as an antifertility agent. In a trial with 181 women in

2,971 cycles (Sakiz et al. 1976), R 2323 administered in a dose of 2.5 mg at weekly intervals believe an index of 7.3% in endometrium biopsy, of which 4.4% was reported as drug failure.

Further studies have been carried out to improve the efficacy of this antiprogestin (Sakiz et al. 1974; Azadian-Boulanger et al. 1976). At the time of ovulation the formation of progesterone receptors is induced by estrogen secretion, at a time when progesterone levels are still low. This period represents the optimum time for an antiprogestin action, as the number of unoccupied progesterone receptors is at its maximum. A dose of 50 mg R 2323 was administered on each of days 5, 16, and 17 of the cycle. This amount of the compound represents the maximum tolerated dose. Nevertheless, a pearl index of 9.5% with 5% drug failure was recorded in 2,148 cycles of 160 women. The use of antiprogestins gives good control of the cycle and few menstrual irregularities occur. In some women 59 mg provoked vomiting and vertigo. Endometrial biopsies performed in the luteal phase showed only a slightly secretory endometrium.

In addition to oral administration, R 2323 can be administered in the form of vaginal rings. In 14 healthy volunteers (Viinikka et al. 1975) vaginal rings with 10, 50, or 200 mg R 2323 were inserted on the first day of menstrual bleeding. This study was performed exclusively to investigate the pharmacokinetics of the drug.

5.3.2 RU 486

Biochemistry. RU 486 is a 19-norsteroid. Its substitutions at positions C-17 and C-11 are reminiscent on the one hand of a progestin, but on the other hand in its overall structure it is similar to antiestrogens of the triphenyl series.

The compound RU 486 binds with high affinity to the progesterone receptor, but lacks progestin-like activity. On the contrary, it acts as an antagonist to progesterone and can thereby inhibit early pregnancy in the luteal phase.

Animal Experiments. The following observations were made in laboratory animals: the affinity of RU 486 for the progesterone receptor in the rat uterus is five times greater than that of progesterone, and its affinity for the glucocorticoid receptor of the rat thymus is three times greater than that of dexamethasone. The affinity for the androgen receptor, in contrast, is low and is negligible for the estrogen and mineralocorticoid receptor, as has been observed by conventional methods.

Other observations made in laboratory animals are: (a) in any stage of pregnancy RU 486 exerts an abortifacient effect; (b) RU 486 has a strong antiglucocorticoid activity without agonistic effect, as demonstrated by in vivo and in vitro experiments in various species (Philibert et al. 1981); (c) there is no estrogen, antiestrogen, mineralocorticoid, or antimineralocorticoid activitiy at the doses tested; (d) the use of RU 486 over a period of 30 days in Sprague–Dawley rats and *Macaca fascicularis* monkeys produced no toxic effects, with the exception of symptoms that can primarily be attributed to the antiglucocorticoid effect at high doses (100 mg/kg per day and more).

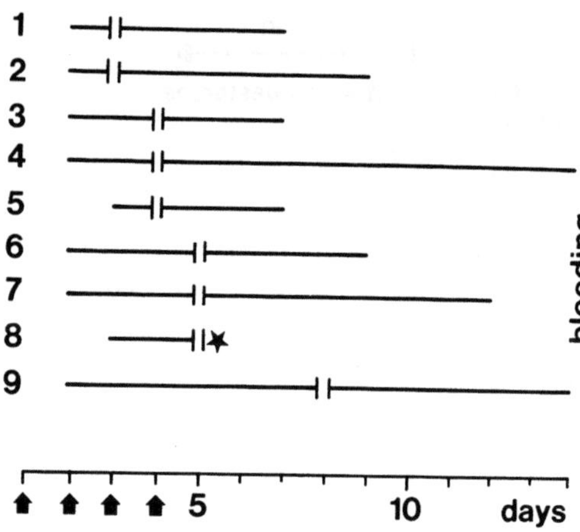

Fig. 7. Clinical trial of RU 486 in nine healthy female volunteers at 6–8 weeks of gestation, with 400 mg daily for 4 days (↑). The beginning of each *line* shows the onset of bleeding and the length indicates the duration of bleeding. The *vertical double line* interrupting each *horizontal line* indicates the day of abortion. (Herrmann et al. 1982)

Clinical Trials. The antiprogestin RU 486 binds with high affinity to the uterine progesterone receptor. In women, practically no affinity for transcortin has been observed.

Herrmann et al. (1982) analyzed the abortifacient effect of RU486 in 11 women volunteers (18–34 years of age), who were 6–8 weeks pregnant and had requested abortion in conformity with the laws of the canton of Geneva. They were given complete information on the experiment before receiving 200 mg RU 486 a day for 4 days, divided into two or four daily doses. In 7 cases, vaginal bleeding occurred on day 1 after administration of the first doses, and in 2 cases, on day 2 (Fig. 7).

Expulsion took place in 2 cases on the 3rd day, in 3 cases on the 4th day, in 3 cases on the 5th day, and in 1 case on the 8th day. Most of the patients experienced some transient lightheadedness (similar to the symptomatology of vaginal stimulation), without further consequences, on the 2nd and 3rd days. This was generally accompanied by uterine cramps. However, blood pressure, serum electrolytes, and organic compounds assayed in the plasma remained constant. Measurement of β-hCG in all cases revealed that the increase common during pregnancy diminishes after expulsion of the conceptus. The same applies to the serum progesterone level, also measured at all stages of observation, which remained constant until expulsion and then diminished rapidly. The estradiol level, when determined, followed the same pattern (Fig. 8).

In contrast, an increase was recorded at 24 h in the serum cortisol level, which quickly returned to normal after treatment had ended. Bleeding associated with abortion lasted for 2–4 days. Spotting continued for another 2–6 days in 4 cases. This general picture is similar to observations made after spontaneous abortion (Fig. 7). Ovarian activity returned to normal in all cases where oral contraception was not administered immediately after treatment. In one case (patient No. 8; Fig. 7) extensive bleeding necessitated a blood transfusion and curettage of the uterus.

In two patients, considered as failures in this experiment, pregnancy was not terminated. They experienced only light bleeding, which began later than in the

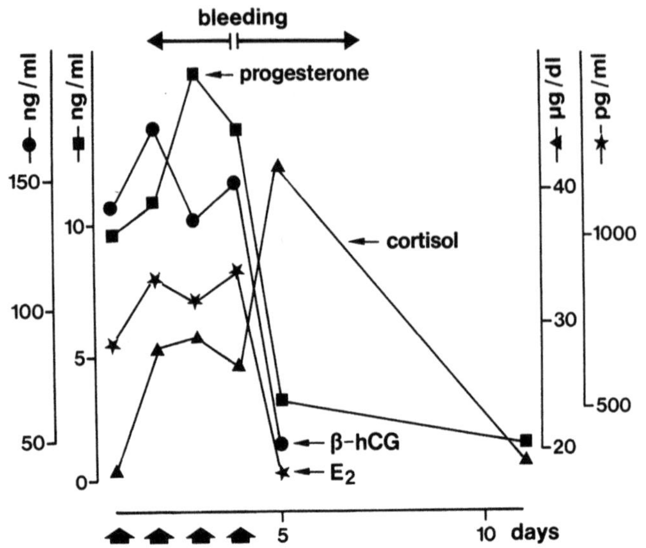

Fig. 8. Effect of four 200-mg doses of RU 486 (↑) to a patient 7 weeks pregnant. The *horizontal line* at the top of the graph indicates the duration of bleeding and the *double vertical line*, abortion. (Herrmann et al. 1982)

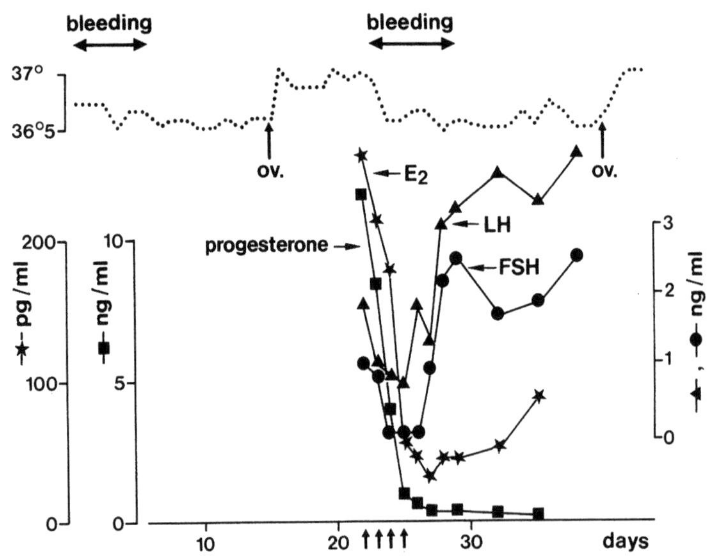

Fig. 9. Interruption of the menstrual cycle on day 22 by administration of 50 mg RU 486 daily for 4 days (↑). The *horizontal lines* indicate duration of bleeding in each case. (Herrmann et al. 1982)

other cases and was not followed by expulsion. No explanation has yet been found. Pregnancy was terminated by aspiration.

Three further female volunteers aged 23–36 years with normal menstrual cycles (all with intrauterine devices) were given 50 mg RU 486 in the morning of day 22 of their menstrual cycle (7 days after ovulation) (Fig. 9). This dose was repeated on each of 3 more days. At 48 h after the first dose bleeding occurred, which was comparable with or heavier than normal menstrual bleedings. Following the first dose of the compound serum progesterone, estrogen, FSH, and LH levels were diminished, as was basal temperature. The next cycle was normal without any side-effects (Herrmann et al. 1982).

5.3.3 RMI 12,936

Biochemistry. A remarkable antiprogestin activity has been attributed to RMI 12,936, which is a Δ^5-3 ketosteroid.

The antifertility effect of RMI 12,936 is due to the enhanced transport of the ovum with its expulsion from the reproductive tract.

Animal Experiments. Preliminary experiments with this compound showed that it could terminate pregnancy in healthy rats when administered before implantation, i.e., on days 1–8 of pregnancy, or shortly before delivery on day 19. RMI 12,936 had similar characteristics to the standard synthetic estrogen, ethinylestradiol, in functional tests. The termination of pregnancy following drug administration on day 1 of pregnancy was linked with a significant simultaneous reduction of ovarian weight, which was postulated to be a cause of corpus luteum regression. Experiments with the transport of the ovum demonstrated a reduced viability of the ova following treatment with RMI 12,936 before implantation. A further important effect is the receptivity of the uterus for the ova.

When RMI 12,936 was given on day 8 of pregnancy resorption of the fetus resulted. This effect occurred together with a significant increase in weight, which was probably caused by luteal hypertrophy. Even in the presence of progesterone implants that were adequate to support pregnancy in ovariectomized rats this effect was irreversible.

Studies to elucidate the mechanisms of antifertility activity in rats (Kendle 1979) showed that luteal hypertrophy followed the administration of RMI 12,936 in pregnant rats but not in pregnant and ovariectomized rats. A similar hypertrophy was achieved in pseudopregnant and immature, gonadotropin-treated rats. Therefore, it seems likely that a direct action on the functional corpus luteum is mediated via a luteotrophic factor of the uterus.

As RMI 12,936 is a Δ^5-3 ketosteroid, it does not compete with labeled progesterone at the cytosolic receptor. One mechanism of action postulated for it is suppression of progesterone biosynthesis. RMI 12,936 may thereby act as a competitive substrate of the Δ^5-3 ketosteroid isomerase, decreasing the isomerization of pregn-5-en-3,20-dione and it is itself isomerated to 7α-methyltestosterone. The 7α-methyltestosterone is either a competitive inhibitor of progesterone on the receptor or is further metabolized. In experiments testing antifertility in rats, 7α-methyltestosterone had a similar effect and potency as RMI 12,936.

Clinical Trials. No data are available on clinical trials with this compound. The results of the experiments conducted with RMI 12,936 with a view to suppression of ovulation for the development of antiprogestins for use in human fertility control suggest that this compound has long-term biological effects following a single administration and that it can act as an antiestrogen as well as an antiprogestin.

Since the original aim of antiprogestin treatment was to influence the cycle for only a short time, long-term effects of RMI 12,936 are undesirable. Further experiments are necessary, to test whether these are due to slow excretion of the parent substance, formation of an active metabolite with a prolonged duration of action, or interruption of the sequence of events, which cannot be induced without an exogenous stimulus.

6 Conclusions

The main application of antiprogestins is for the pharmacological control of human fertility. Since 1962 there have been reports (Banik and Pinkus 1962) on specific antiprogestins. In spite of remarkable advances in this field no compound has provided the ideal combination of high potency and high specificity. Instead of a complete review of the literature, this report deals with the current studies concerned with the development of improved compounds approaching an ideal substance.

According to the previously mentioned aims, antiprogestins have been developed that compete with endogenous progesterone at its specific receptors and at the same time lack premenstrual activity, i.e., have no progesterone-agonistic effects. The mechanism of the antiprogestins on the human endometrium inhibits the implantation of the fertilized ovum at midcycle. Various antiprogestins with irreversible covalent binding to the receptor have already been described, which, however, cannot be applied clinically owing to high toxicity. Recently, further nontoxic compounds have been developed. These substances are intended to occupy the free progesterone receptors in the late proliferative phase. One of these compounds, 16α-bromoacetoxyprogesterone, is an alkalizing agent, but it may be carcinogenic.

More novel antiprogestins, such as RU 486, are powerful abortifacients without any mechanical action. They have similar disadvantages of older antiprogestins, in particular with respect to bleeding. These early abortifacients are more effective when administered as early as possible. Antiprogestins can also be used as cycle regulators to provide regular 28-day cycles. The appropriate substances can be administered at midcycle in the early luteal phase before or after the expected menstruation, whereby the number of free receptor sites in the endometrium will be occupied during the late proliferative phase and the secretory proliferation of the endometrium will be inhibited. The site of action of antiprogestins is either the human endometrium or the myometrium. This means that bleeding will occur either as a normal menstrual period that would have taken place anyway or due to prevention of the implantation of an undiscovered conceptus.

The safety of antigestagens for use in contraception is increased if the pharmacological disturbance of the menstrual cycle can be restricted to 1 or 2 days within it. A method that depends on administration at a precisely determined point in the cycle will lose some of its efficiency, since the interdependent events of the menstrual cycle vary among women and with each cycle. Therefore, the ideal compound must be sufficiently effective regardless of the exact time of administration within the cycle. In spite of the difficulty of developing such an ideal compound this approach to finding an improved method of contraception looks very promising.

Other investigations on antiprogestins are possible: study of pituitary–ovarian interaction, regulation of placenta function, fallopian tube preservation following extrauterine pregnancy, treatment of tumors (deciduomas, breast cancer), the role of progesterone in the mechanism of childbirth, and the adrenal function of hypothalamic-pituitary control.

In summary, antiprogestins seem to potentially useful compounds for menstrual regulation, contraception, or induction of abortion. Further clinical trials will be necessary to clarify the optimum doses and rhythms and the effects of long-term use.

Acknowledgement. We wish to thank Prof. Etienne-Emile Baulieu (Bicetre, France) for reviewing our manuscript.

References

Aitken RJ, Harper MJK (1977) New methods for the regulation of implantation. Contraception 16:227–241

Anderson J, Clark JH, Peck EJ (1972) Oestrogen and nuclear binding sites: determination of specific sites by ^{3}H-oestradiol exchanges. Biochem J 126:561–567

Atger M, Baulieu E-E, Milgrom E (1974) An investigation of progesterone receptors in guinea pig vagina, uteine cervix, mammary glands, pituitary and hypothalamus. Endocrinology 94:161–167

Azadian-Boulanger G, Secchi J, Laraque F, Raynaud JP, Sakiz E (1976) Action of a midcycle contraceptive (R 2323) on the human endometrium. Am J Obstet Gynecol 125:1049–1056

Bacic M, Wesselius de Casparis A, Diczfalusy E (1970) Failure of large doses of ethinyl estradiol to interfere with early embryonic development in the human species. Am J Obstet Gynecol 107:531–534

Bahl OP (1969) Human chorionic gonadotrophin II. Nature of the carbohydrate units. J Biol Chem 244:575–583

Banik UD, Pincus G (1962) Effect of steroidal antiprogestins on implantation of fertilised eggs of rats and mice. Proc Soc Exp Biol Med 111:595–602

Baulieu E-E, Alberga A, Jung I, Lebeau M-C, Mercier-Bodard C, Milgrom E, Raynaud J-P, Raynaud-Jammet C, Rochefort H, Truong H, Robel P (1971) Metabolism and protein binding of sex steroids in target organs: an approach to the mechanism of hormone action. Recent Prog Hormone Res 27:351

Bayard F, Damilano S, Robel P, Baulieu E-E (1975) Récepteurs de l'oestradiol et de la progestérone dans l'endomètre humain au cours du cycle menstruel. C R Séances Acad Sci 281:1341–1344

Beyer G, Zeilmarker GH (1974) Prolonged pseudo-pregnancy in mice bearing ectopic trophoblastic tissue. J Endocrinol 61:509–510

Beyer B, Terenius L, Brueggemeier RW, Ranade VV, Counsell RE (1976) Synthesis of potential antiprogestins. Steroids 27:123–131
Bouin P, Ancel P (1910) Sur les fonctions du corps jaune gestatif. I. Sur le déterminisme de la préparation de l'utérus à la fixation de l'oeuf. J Physiol Pathol Gen 12:1–16
Brown WE, Bradbury JT (1947) A study of the physiologic action of human chorionic hormone. Am J Obstet Gynecol 53:749–757
Butenandt A, Westphal U, Hohlweg W (1934) Über das Hormon des Corpus lutenum. Hoppe–Zeylers Z Physiol 227:84–98
Chamboon Y (1949) Besoins endocriniens qualitatifs et quantitatifs de l'ovoimplantation chez la lapine. C R Soc Biol (Paris) 143:1172–1175
Channing CP, Sakai C, Bahl OP (1976) Role of the carbohydrate residues of human chorionic gonadotropin (hCG) on its ability to bind and stimulate cyclic AMP accumulation in porcine granulosa cells. Fed Proc 35:798
Channing CP, Sakai C, Bahl OP (1977) Role of carbohydrate residues of human chorionic gonadotropin in binding and stimulation of cyclic AMP and progesterone secretion by porcine granulosa cells. Endocrinology 103:341–348
Channing CP, Engel B, Bahl OP (1978) Role of carbohydrate residues of human chorionic gonadotropin in stimulation of luteinization of monkey granulosa cell cultures. Biol Reprod 18:707
Clark SW, Sweet F, Warren JC (1974) Interceptive activity of 16α-Bromoacetoxyprogesterone. Biol Reprod 11:519–528
Clark SW, Sweet F, Warren JC (1975) Synthesis and use of affinity-labeling steroids for interceptive purposes. Am J Obstet Gynecol 121:864
Courrier R (1945) cited in: Reynolds SRM, Hoeber PH (eds) Physiology of the uterus. Harper New York (1949) p 111
Csapo AI (1976) Prostaglandin impact. In: Samuelsson B, Paoletti R (eds) Advances in prostaglandin and thromboxane research, vol 2. Raven, New York, pp 705–718
Csapo AI, Pulkkinen MP (1973) The effect of estradiol replacement therapy on early pregnant luteectomized patients. Am J Obstet Gynecol 117:987–990
Csapo AI, Pulkkinen M (1978) Indispensability of human corpus luteum in the maintenance of early pregnancy luteectomy evidence. Obstet Gynecol Survey 83:69–81
Csapo AI, Pulkkinen MP, Kaihola HL (1974) The relationship between timing of luteectomy and the incidence of complete abortions. Am J Obstet Gynecol 118:985–989
Deanesly R (1960) Implantation and early pregnancy in ovariectomized guinea-pigs. J Reprod Fertil I:242
Feder HH, Resko JA, Goy RW (1968) Progesterone concentrations in the arterial plasma of guinea-pigs during the oestrous cycle. J Endocrinol 40:505
Fraenkel S (1910) Die Funktion des Corpus luteum. Arch Gynäkol 68:438
Fried PH, Rakoff AE (1952) The effects of chorionic gonadotropin and prolactin on the maintenance of corpus luteum function. Eur J Clin Endocrinol Metabol 12:321–327
Fuchs AR, Beling CG (1974) Evidence of early ovarian recognition of blastocysts in rabbits. Endocrinology 95:1054–1058
Gase JM, Renoir JM, Radanyi CH, Joab I, Tuohima P, Baulieu EE (1984) Progesterone receptor in the chick oviduct: an immunohistochemical study with antibodies to distinct receptor components. J Cell Biol (in press)
Good RG, Moyer DL (1968) Estrogen-progesterone relationships in the development of secretory endometrium. Fertil Steril 19:37–49
Guériguian JL, Sawyer ME, Pearlman WH (1974) A comparative study of progesterone- and cortisol-binding activity in the uterus and serum of pregnant and non-pregnant women. J Endocrinol 61:331–345
Hanson FW, Powell JE, Stevens VC (1971) Effects of hCG and human pituitary LH on steroid secretion and functional life span of the human corpus luteum. J Clin Endocrinol Metabol 32:211–215
Haour F (1976) Rat chorionic gonadotropin (rCG): radioreceptor assay and correlation with corpus luteum function during gestation. In: V International Congress of Endocrinology Excerpta Medica Int Congr Series, Amsterdam (Abstract 777), p 321

Haukkamaa M, Karjalainen O, Luukkainen T (1971) In vitro binding of progesterone by the human endometrium during the menstrual cycle and by hyperplastic, atrophic, and carcinomatous endometrium. Am J Obstet Gynecol 111:205
Henzl MR, Segre EJ, Nakamura RM (1973) The influence of oxymetholone on the HCG-stimulated corpus luteum. Contraception 8:515–520
Herrmann W, Wyss R, Riondel A, Philibert D, Teutsch G, Sakiz E, Baulieu E-E (1982) Effect of an antiprogesterone in women: interruption of the menstrual cycle and of early pregnancy. C R Séances Acad Sci 294:933–938
Jaffe RB, Lee PA, Midgley AR (1969) Serum gonadotropins before, at the inceptions of, and following human pregnancy. J Clin Endocrinol Metab 29:1281–1283
Jänne O, Kontula K, Luukkainen T, Vihko R (1975) Oestrogen-induced progsterone in human uterus. J Steroid Biochem 6:501–509
Johannson DB (1971) Depression of the progesterone levels in women treated with synthetic gestagens after ovulation. Acta Endocrinol (Copenh) 68:779
Kendle KE (1979) Current investigations of antiprogestational steroids. In: Agrawal MK (ed) Antihormones. Elsevier/North Holland Biomedical Press Amsterdam, pp 293–305
Knobil E (1973) On the regulation of the primate corpus luteum. Biol Reprod 8:246–258
Kontula K (1975a) Cytosol progesterone receptor in mammalian uterus. Academic Dissertation, University of Helsinki, Finland
Kontula K (1975b) Progesterone-binding proteins from endometrium and myometrium of sheep uterus: a comparative study. Acta Endocrinol (Copenh) 78:593–603
Kontula K (1975c) Progesterone binding protein in human myometrium binding site concentration in relation to endogenous progesterone and estradiol-17β levels. J Steroid Biochem 6:1555–1561
Kontula K, Jänne O, Luukkainen T, Vihko R (1973) Progesterone-binding protein in human myometrium. Ligand specificity and some physicochemical characteristics. Biochem Biophys Acta 328:145–153
Kontula K, Jänne O, Luukkainen R, Vihko R (1974a) Progesterone-binding protein in human myometrium. Influence of metal ions on binding. J Clin Endocrinol Metabol 38:500–503
Kontula K, Jänne O, Rajakoski E, Tanhuanpää E, Vihko R (1974b) Ligand specificity of progesterone-binding proteins in guinea pig and sheep. J Steroid Biochem 5:39–44
Kontula K, Jänne O, Vihko R, Jager E, Visser J, Zeelen F (1975) Progesterone-binding proteins: in vitro binding and biological activity of different steroidal ligands. Acta Endocrinol (Copenh) 78:574–592
Leavitt WW, Toft DO, Strott CA, O'Malley BW (1974) A specific progesterone receptor in the hamster uterus: physiologic properties and regulation during the estrous cycle. Endocrinology 94:1041–1053
Luu Thi MT, Baulieu E-E, Milgrom E (1975) Comparison of the characteristics and of the hormonal control of endometrial and myometrial progesterone receptors. J Endocrinol 66:349–356
McGuire JL, De Della C (1971) In vitro evidence for a progestogen receptor in the rat and rabbit uterus. Endocrinology 88:1099–1103
Meyer RK (1972) Chorionic gonadotropin, corpus luteum function and embryo implantation in the rhesus monkey. In: Dicfalusy E, Standley CC (eds) The use of non-human primates in research on human reproduction. WHO, Genf, pp 214–217
Milgrom E, Baulieu E-E (1970a) Progesterone in uterus and plasma. I. Binding in rat uterus 105,000 g supernatant. Endocrinology 87:276–287
Milgrom E, Baulieu E-E (1970b) Progesterone in the uterus and the plasma. II. The role of hormone availability and metabolism on selective binding to uterus protein. Biochem Biophys Res Commun 40:723–730
Milgrom E, Atger M, Baulieu E-E (1970) Progesterone in uterus and plasma. IV. Progesterone receptor(s) in guinea pig uterus cytosol. Steroids 16:741–754
Milgrom E, Perrot M, Atger M, Baulieu E-E (1972a) Progesterone in uterus and plasma. V. An assay of the progesterone Cytosol receptor of the guinea pig uterus. Endocrinology 90:1064–1070

Milgrom E, Atger M, Perrot M, Baulieu E-E (1972b) Progesterone in uterus and plasma. Uterine progesterone receptors during the estrus cycle and implantation in the guinea pig. Endocrinology 90:1071–1078

Milgrom E, Luu Thi M, Atger M, Baulieu E-E (1973a) Mechanism regulating the concentration and the conformation of progesterone receptor(s) in the uterus. J Biol Chem 248:6366–6374

Milgrom E, Luu Thi MT, Baulieu E-E (1973b) Control mechanisms of steroid hormone receptors in the reproductive tract. Acta Endocrinol (Copenh) [Suppl] 180:380–403

Neill JD, Johansson EDB, Knobil E (1969) Failure of hysterectomy to influence the normal pattern of cyclic progesterone secretion in the rhesus monkey. Endocrinology 84:464–465

Neumann F, Elger W (1971) Kritische Überlegungen zu den biologischen Grundlagen von Toxizitätsstudien mit Seroid-(Sexual-)hormonen. In: Plotz EJ, Haller J (eds) Methodik der Steroidtoxikologie. Georg Thieme, Stuttgart, pp 6–48

Niswender GD, Menon KMJ, Jaffe RB (1972) Regulation of the corpus luteum during the menstrual cycle and early pregnancy. Fertil Steril 23:432–442

Oriol-Bosch A, Cortes J (1975) Effects of postovulatory estradiol benzoate administration on womens' ovarian function. Fertil Steril 26:405–412

Philibert D, Raynaud JP (1974) Binding of progesterone and R 5020, a highly potent progestin, to human endometrium and myometrium. Contraception 10:457–466

Philibert D, Dereadt R, Teutsch G (1981) 8th International Congress of Pharmacology, Tokyo, Congress Proceedings, Abstract No. 1463, p g668

Rao BR, Wiest WG, Allen WM (1973) Progesterone "receptor" in rabbit uterus. I. Characterization and estradiol-17β augmentation. Endocrinology 92:1229–1240

Raynaud JP, Bonne C, Bouton MM, Moguilewsky M, Philibert D, Azadian-Boulanger G (1975) Screening for anti-hormones by receptor studies. J Steroid Biochem 6:615–622

Raynaud JR, Ojasoo T, Labrie F (1981) Steroid hormones – agonists and antagonists. In: Lewis GP, Ginsburg M (eds) Mechanism of steroid action. Macmillan, London, pp 145–158

Reel JR, Humphrey RR, Shih Y, Windsor B, Sakowski R, Greger PL, Edgren R (1979) Competitive progesterone antagonists: receptor binding and biologic activity of testosterone and 19-nortestosterone derivates. Fertil Steril 31:553

Reinius S, Fritz GR, Knobil E (1973) Ultrastructure and endrocrinological correlation of an early implantation site in the rhesus monkey. J Reprod Fertil 32:171–173

Ross GT, Cargille GM, Lipsett MB, Rayford PL, Marshall JR, Strott CA, Rodbard D (1970) Pituitary and gonaldal hormones in women during spontaneous and induced ovulatory cycles. Recent Prog Hormone Res 26:1

Sakai CN, Engel B, Channing CP (1977) Ability of extract of pig corpus luteum to inhibit binding of ^{125}I-labelled human chorionic gonadotropin in porcine granulosa cells. Proc Soc Exp Biol 155:373–376

Sakiz E, Azadian-Boulanger G (1971) R 2323 – an original contraceptive compound. In: James VHT, Martini L (eds) Proc IIIrd Int Congr Hormonal Steroids, Hamburg 1970. Excerpta Medica Int Congr Series 219:865–871

Sakiz E, Azadian-Boulanger G, Raynaud JP (1972) Antiestrogens, antiprogesterones. In: Proceedings of the Fourth International Congress of Endocrinology, Washington. Excerpta Medica, Int Congr Series, Amsterdam, 18–24 June 1972, pp 988–994

Sakiz E, Azadian-Boulanger G, Larraque F, Raynaud JP (1974) A new approach to estrogen-free contraception based on progesterone receptor blockade by mid-cycle administration of ethyl norgestrienone (R 2323). Contraception 10:467–480

Sakiz E, Azadian-Boulanger G, Ojasoo T, Laraque F (1976) Contraceptive efficacy of one-weekly oral administration of 2,5 mg R 2323. Contraception 14:275–284

Saxena BB, Hasan SH, Haour F, Schmidt-Gollwitzer M (1974) Radioreceptor assay of human chorionic gonadotropin: detection of early pregnancy. Science 184:793–795

Short RV (1969) Implantation and the maternal recognition of pregnancy. In: Wolstenholme GEW, O'Connor M (eds) Ciba Symposium on Foetal Autonomy. Churchill, London, pp 2–31

Short RV (1972) Role of hormones in sex cycles. In: Austin CR, Short RV (eds) Reproduction in mammals, vol 3. University Press, Cambridge, p 42
Smith RG, Iramain CA, Buttram VC, O'Malley BW (1975) Purification of human uterine progesterone receptor. Nature 253:271–272
Thomas PJ (1973) Steroid hormones and their receptors. J Endocrinol 57:333–359
Trams G, Brewitt H, Möllmann G, Maas H (1973) Effect of progesterone on RNA and protein synthesis in the rat uterus. Acta Endocrinol (Copenh) 73:740–750
Van de Wiele RL, Bogumil J, Dyrenfurth I, Ferin M, Jewelewicz R, Warren M, Rizkallah T, Mikhail G (1970) Mechanisms regulating the menstrual cycle in women. Recent Prog Hormone Res 26:63
Viinikka L, Victor A, Jänne O, Raynaud J-P (1975) The plasma concentration of a synthetic progestin, R 2323, released from polysilastic vaginal rings. Contraception 12:309–316
Warren JC (1973) Progesterone: implications for fertility control. Biol Reprod 8:259
Wiest WG, Rao BR (1971) Progesterone binding proteins in rabbit uterus and human endometrium. Schering Workshop on steroid hormone "receptors", Berlin 7–9 December 1970. In: Raspé G (ed) Advances in the biosciences, vol 7. Pergamon, Oxford, p 251

Progesterone Receptors in Brain and Hypophysis

JUNZO KATO

Contents

1 Department of Obstetrics and Gynecology, Yamanashi Medical University, Tamaho, Nakakoma-gun, Yamanashi, 409-38 Japan

Current Topics in Neuroendocrinology, Vol 5

1 Introduction

It is well established that progesterone plays a role in the brain and hypophysis as a facilitator and inhibitor of sexual behavior and gonadotropin release in the female rat (Everett 1961; Caligaris et al. 1971; Brown-Grant and Naftolin 1972; Dörner 1972; Meyerson 1972; Barraclough 1973; Goldman and Zarrow 1973; Mann and Barraclough 1973; Freeman et al. 1976; Feder and Marrone 1977; Goodman 1978; Attardi 1981), guinea pig (Morin and Feder 1974), and primates (Odell and Swerdloff 1968; Spies and Niswender 1972; Yamaji et al. 1972; Karsch et al. 1973; Dierschke et al. 1973; Knobil 1974; Clifton et al. 1975).

In an attempt to learn whether a specific progesterone uptake mechanism exists in the brain and the hypophysis, the distribution and retention pattern of radioactivity after in vivo injection of labeled progesterone was studied. Early work of Kato (1963) did not show a selective uptake of radioactivity in the hypothalamus of immature and estrogen-primed immature rats after injection of low-specific-activity [^{14}C]progesterone, but some tendency of the reticular formation to take up radiation was observed. Laumas and Farooq (1966) reported that after intravenous administration of labeled progesterone to ovariectomized estrogen-treated rats, radioactivity in the brain and pituitary appeared to show a very slight, insignificant increase 1–2 min after injection, but the uptake pattern was not definite, as had been seen with estradiol. Seiki et al. (1968, 1969) carried out similar experiments on adult ovariectomized rats killed 2–60 min after intravenous injection of [^{3}H]progesterone. There was no evidence of specific progesterone uptake in the hypothalamus and hypophysis, although there was a tendency for progesterone to accumulate in the posterior hypothalamus at a higher rate than in the other parts. Further attempts to obtain unequivocal evidence for the presence of specific progestin binding in areas of the brain were for the most part unsuccessful (Wade and Feder 1972; Whalen and Gorzalka 1974; Reel and Shin 1975; McEwen et al. 1976; Marrone and Feder 1977; Rahman et al. 1977). In contrast, there was evidence of progesterone uptake by some specific areas of the brain. Seiki and Hattori (1971) reported that the median eminence and hypophysis were able to concentrate labeled progesterone in vivo. Whalen and Luttge (1971 a, b) suggested the existence of limited-capacity binding sites for progesterone in the hypothalamus and hypophysis of the female rat. Neurons of the guinea pig hypothalamus had been shown autoradiographically to concentrate ^{3}H-progesterone or its metabolites (Sar and Stumpf 1973; Warembourg 1978 a).

The evidence for the identification of specific progesterone receptor proteins in the brain and hypophysis has been both controversial and conflicting. Seiki and Hattori (1973) and Seiki et al. (1977) reported the presence of progesterone-binding proteins of 4.3*S* and 7.1*S* in the cytosol of both tissues. These proteins have a limited capacity for binding steroids, as shown by gel filtration on Sephadex G-100. Iramain et al. (1973), using the charcoal-dextran technique, also reported high-affinity–low-capacity progesterone binding in hypothalamic and hypophyseal cytosols of female guinea pigs and rabbits, although they were unable to demonstrate peaks of progesterone binding by sucrose gradient analysis. However, Atger et al. (1974) reported an absence of receptors in the hypothalamus of castrated estrogen-primed guinea pigs. Kato (1975) was also unsuccessful in identifying specific 7*S* ^{3}H-progesterone binding in rat hypothalamic cytosol on sucrose gradients, in sharp contrast to the presence of a large 7*S* peak in the uterine cytosol.

The inconsistent results of the above-mentioned reports in which ^{3}H-progesterone was used as radioactive ligand might be explained by the following: Since progesterone binds to receptors for progesterone and corticoids, and to corticosterone-binding globulin (CGB), it is difficult to differentiate specific progesterone receptors from contaminating CBG, CBG-like progesterone-binding pro-

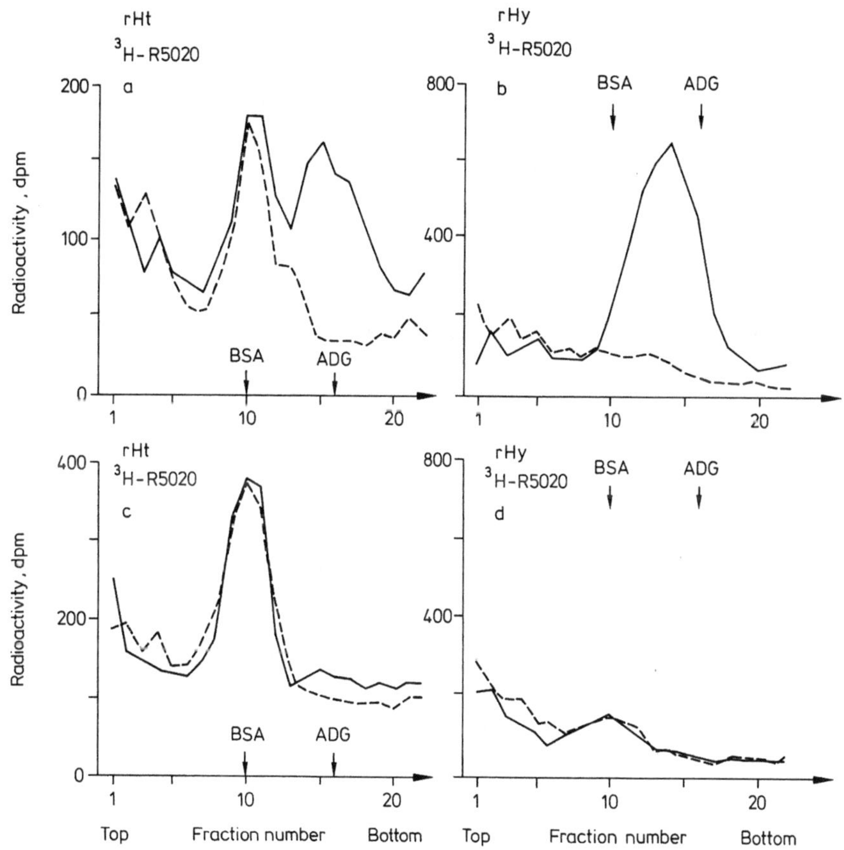

Fig. 1 a–d. Representative density gradient profiles of progestin-binding receptors in hypothalamic and hypophyseal cytosols from nonprimed and estrogen-primed 27-day-old female rats. Estrogen-primed rats were injected subcutaneously with 5 μg of estradiol benzoate once daily for 7 consecutive days and then killed. Cytosols (0.29 ml) from the hypothalamus (3.0 mg protein) und hypophysis (1.0 mg protein) were incubated with ^{3}H-R5020 in the presence or absence of excess unlabeld R5020. The mixture was layered on 5%–20% sucrose density gradients containing 10% glycerol, followed by centrifugation at 50000 rpm in a SW50.1 rotor with a Beckman ultracentrifuge for 18 h. **a** Hypothalamus, estrogen-primed (^{3}H-R5020, 3.7 n*M*); **b** in Hypophysis, estrogen-primed (7.7 n*M*); **c** Hypothalamus, nonprimed (7.7 n*M*); **d** Hypophysis, nonprimed (7.7 n*M*). [From Kato and Onouchi (1977)]

teins, or corticoid-binding proteins; secondly, the progesterone-receptor complexes are unstable; and thirdly the brain contains only a small amount of the receptors (Philibert and Raynaud 1973; Kato et al. 1978).

The use of ^{3}H-R5020 as ligand has made it possible to overcome the disadvantages met with in using labeled progesterone. R5020, a synthetic progestin (17,21-dimethyl-19-nor-4,9-pregnadiene-3,20-dione; Roussel-Uclaf, Romainville, France) has been shown to specifically bind to progestin receptors, with slower dissociation of ligand-receptor complexes and higher affinity than labeled progesterone (Philibert and Raynaud 1973). With ^{3}H-R5020, the presence of high-affinity–low-capacity specific progestin-binding macromolecules has been clearly demonstrated in hypothalamic and hypophyseal cytosols from the estrogen-primed rat (Kato 1977a; Kato and Onouchi 1977; Moguileswky and Raynaud 1977, 1979a; Kato et al. 1978; MacLusky and McEwen 1978). A possible role for these macromolecules was suggested to be in the mechanism of action of progesterone on the brain. Figure 1 shows representative sedimentation profiles of progestin receptors in hypothalamic and hypophyseal cytosols from estrogen-primed immature (28-day-old) female rats.

2 Identification and Characterization of Cytosolic and Nuclear Progestin Receptors in the Brain and Hypophysis

2.1 Rodents

2.1.1 Cytosolic Progestin Receptors in the Brain and Hypophysis

High-affinity–low-capacity progestin-binding macromolecules have been identified in hypothalamic and hypophyseal cytosol from the rat (Kato and Onouchi 1977; Seiki et al. 1977; Moguilewsky and Raynaud 1977, 1979a; Greenstein 1978; MacLusky and McEwen 1978, 1980a; Lee et al. 1979), Hamster (Leavitt et al. 1977), guinea pig (Blaustein and Feder 1979), and monkey (MacLusky et al. 1980; Kato et al. 1981, this chapter); as well as in the hen (Kawashima et al. 1978) and male ring dove (Balthazart et al. 1980).

2.1.1.1 Characterization of Cytosolic Progestin Receptors

The physiochemical characterizations of brain and hypophyseal receptors are summarized in Table 1.

The dissociation constants (K_d) of the receptors in these tissues are in the range of 0.1–3.3 nM. The number of binding sites in the hypothalamus und hypophysis range from 10 to 63 and from 31 to 599 fmol/mg cytosol protein, respectively. The hypothalamus and hypophysis of birds seem to contain greater concentrations of the receptors than those of mammals (Table 1). The progestin-receptor complexes sediment in the 6S–7S region on sucrose gradients (Table 1). The brain and hypophyseal receptors are, in part at least, proteinaceous (Kato and Onouchi 1977).

Table 1. Properties of cytosolic progestin receptors in the brain and hypophysis[a]

Kimals	Sedimentation coefficient (*S*)	Dissociation constant (*Kd*) (nM)	Number of binding sites (fmol/mg protein)	References
at				
Kmmature(28-do), f.	H; 7	H; 3.3 ± 0.8^{b} (SD)	10 ± 1.9 (SD)	Kato and Onouchi (1977)
Adult; OVX	H; 7	–	–	
	P; 7	P; 3.8 ± 0.23 (SD)	31 ± 5.9 (SE)	
Adult; OVX; H, ^{3}H-P				Seiki et al. (1977)
Kmmature, f.; H; P	–	$k_{-1} - 1 \times 10^{-2} min^{-1}$	–	Moguilewsky and Raynaud (1979a)
Adult; OVX; H, P	6–7	H; 1.13 ± 0.09	45 ± 2	Moguilewski and Raynaud (1979b)
	6–7	P; 1.16 ± 0.16	360 ± 16	
Adult, OVX, ^{3}H-P	–	~ 10	H; 135 ± 8 dpm/mg tissue C; 140 ± 10	Greenstein (1978)
Adult, OVX-ADX; H, P	–7 –7	0.26–0.38	H; 36.5 ± 7.5 P; 273 ± 11.8 C; 26.5 ± 3.0	MacLusky and McEwen (1978)
Adult, OVX, f; H, ^{3}H-P	8	$6 \sim 10$	H; 12	Lee et al. (1979)
umster				
Adult, OVX; H, P, ^{3}H-P	6–7	–	P>H	Leavitt et al. (1977)
kinea pig				
DVX	–	H; 0.16 C; 0.24 P; 0.35	H; 33.6 C; 8.7 P; 52.0	Blaustein and Feder (1979a)
onkey				
Bonnet monkey	6–7	0.17–0.26	II; 14–15 P; 61–90	MacLusky et al. (1980)
Rapanese monkey	7	0.1–0.3	H; 15 P; 110	Kato et al. (1981, this chapter)
d				
Hen, ^{3}H-P	–	H; 0.33 ± 0.01^{c} (SE) P; 0.17 ± 0.01 (SE)	43 ± 0.3 (SE) 599 ± 8.4 (SE)	Kawashima et al. (1980)
Ring dove, male, castrated	–	H; 0.4	POAH; 63 ± 4 PH; 33 ± 5	Balthazart et al. (1980)

Abbreviations: H, Hypothalamus and preoptic area; C, cerebral cortex; P, hypophysis; OVX, ovariectomized; f, female; ADX, adrenalectomized; ^{3}H-P, ^{3}H-Progesterone

[a] The animals were treated with estrogen, except for the laying hens

[b] Mean and standard deviation (SD)

[c] Mean and standard error (SE)

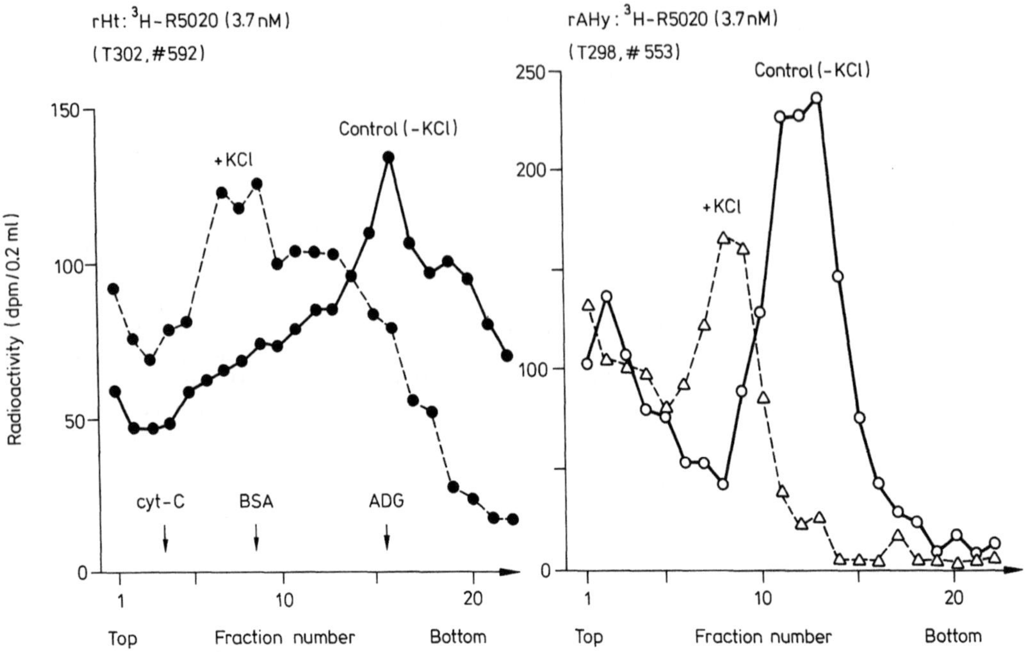

Fig. 2. Density gradient patterns of hypothalamic and anterior hypophyseal progesterone receptors labeled with ^{3}H-R5020. The 7*S* peak of ^{3}H-R5020 binding was collected and incubated in vitro with ^{3}H-R5020, in a medium with or without high salt (0.4 *M* KCl). (—), Control (without KCl); (---), with KCl. [From Kato and Onouchi (1979)]

As shown in Fig. 2 the 7*S* peak appearing in the controls (without KCl treatment) was found to be located in the 4*S* region after KCl treatment (Kato and Onouchi 1979). Hypothalamic and hypophyseal cytosols were first incubated with ^{3}H-R5020 at 0 °C and centrifuged on 5%–20% sucrose density gradients. The concentrated 7*S* fraction was divided into equal volumes, each of which was further incubated in the presence or absence of a final concentration of a 0.4 *M* KCl, followed by centrifugation on a sucrose density gradient, again with or without 0.4 *M* KCl. The progestin receptor macromolecule proteins in the 7*S* region at low ionic strength appear to dissociate into 4*S* subunits at higher ionic strength (0.3–0.4 *M* KCl), suggesting a aggregate-subunit relationship (Kato and Onouchi 1979; MacLusky and McEwen 1980 a).

2.1.1.2 Further Characterization of Cytosolic Progesterone Receptors

Cytosolic Exchange at Low Temperature. Greenstein (1978) has reported that complete exchange occurs in vitro at 0 °C between ^{3}H-progesterone and unlabeled progesterone, and the reaction attains apparent equilibrium within 2 h at this temperature. An incubation at 30 °C resulted in an almost complete loss of saturable binding. Similarly, complete exchange, using rat hypothalamic and hypophyseal progestin receptors, has been seen between ^{3}H-R5020 and unlabeled R5020 or progesterone at low temperatures (0° or 10 °C), but the exchange decreased markedly at 30 °C (Kato and Onouchi 1979). Cytosolic progestin exchange in the brain and hypophysis at low temperatures contrasts with estrogen exchange in uterus, where complete exchange occurs at higher temperatures (25–30 °C) (Katzenellenbogen et al. 1973). The brain and hypophyseal cytosolic progestin receptor seems to be unstable, with rapid degradation at higher temperatures.

Separation on Polyacrylamide Agarose Gel. The progestin receptor proteins in the hypothalamic and hypophyseal cytosols are well separated on polyacrylamide agarose gels from estrogen receptors [labeled with ^{3}H-R2858: moxestrol, 3,17β-dihydroxy-17α-ethynyl-11β-methoxy-estra-1,3,5(10)-triene] and androgen receptors (labeled with ^{3}H-R1881: 17β-hydroxy-17α-methyl-estra-4,9,11-triene-3-one), although they all sediment in the 7*S*–8*S* region on densitiy gradients. This suggests some differences in physicochemical properties between progestin receptor and androgen receptor proteins (Kato and Onouchi 1979).

2.1.2 Nuclear Progestin Receptors

While progestin uptake by brain cell nuclei had been detected autoradiographically in the guinea pig (Sar and Stumpf 1973; Warembourg 1978a) and rat (Warembourg 1978b, c), and by liquid scintillation counting in the rat (Blaustein and Wade 1978), other studies had failed to produce conclusive evidence of saturable progestin binding in cell nuclei of the rodent brain or hypophysis (Atger et al. 1974; McEwen et al. 1976; Marrone and Feder 1977).

On the basis of an in vitro exchange reaction of R5020 and progesterone, saturable binding sites, and isolation of KCl-extractable progestin-binding components by gradient centrifugation, specific progestin receptors have since been demonstrated in nuclei from the hypothalamus and hypophysis of the rat (Kato et al. 1978; Kato and Onouchi 1979; McGinnis et al. 1981b) and the guinea pig (Blaustein and Feder 1980), as well as the hen (Kawashima et al. 1979). The existence of nuclear progesterone receptors, together with the cytosolic progesterone receptors in the hypothalamus and hypophysis, suggests the interaction of progesterone with brain receptors as the basis for the mechanism of its central action.

2.1.2.1 Characterization of Nuclear Progestin Receptors

KCl-Extractable Receptors. The values for sedimentation coefficients (*S*), dissociation constants (*Kd*) and the number of binding sites (NBS) of the KCl-extractable nuclear components are given in Table 2. The components are high-affinity, low-capacity with 5*S*.

Representative sedimentation patterns of ^{3}H-R5020–receptor complexes from 0.4 *M* KCl extract of purified and crude hypothalamic or anterior hypophyseal nuclei are shown in Fig. 3 (Kato and Onouchi 1979). Estradiol valerate-primed adult female rats were s.c. injected with progesterone (2 mg) 1 h before sacrifice. The 800 g pellet and purified nuclei from hypothalami and anterior hypophyses of the rats were incubated at 10 °C for 2 h with ^{3}H-R5020, followed by

Table 2. Characterization of nuclear progestin receptors in the brain and hypophysis

Animals	Sedimentation coefficient (*S*)	Dissociation constant (*Kd*) (nM)	Number of binding sites (NBS) (fmol/mg protein)	References
Rat				
OVX-ADX, EB-primed adult rats In vivo injection of ^{3}H-R 5020	Cell nuclear, saturable progestin-specific binding No data on the values for *S*, *Kd* and NBS			Blaustein and Wade (1978)
Adult, OVX, EB-primed rats 0.4 *M* KCl extracted	H; 5S P; 5S	– –	Very low	Kato et al. (1978)
Immature female, E_2-primed 0.4 *M* KCl extracted	– –	H; ~1.4 P; 6.4	466 915	Kato and Onouchi (1979)
Adult, estradiol-valerate-primed female rats	H; 5S P; 4–5S	– –	– –	
OVX, E_2-primed rats	Nuclear translocation of PR in the HPOA			McGinnis et al. (1981b)
Guinea pig				
In vitro exchange ^{3}H-R 5020	–	0.16	179–189	Blaustein and Feder (1980)
Hypothalamus-preoptic area and septum (HPS)	Nuclear translocation of PR in HPS. No replenishment of PRc.			Blaustein and Feder (1979a) Blaustein and Feder (1979b)
Bird				
Hen (*Gallus domesticus*) ^{3}H-Progesterone Exchange	–	H; 0.29 P; 0.46	290 210	Kawashima et al. (1979)

Abbreviations: H, Hypothalamus; P, Hypophysis; HPOA, Hypothalamus-preoptic area; OVX, ovariectomized; EB, estradiol benzoate; E_2, estradiol; PR, progestin receptor; PRc, cytosolic progestin receptor

extraction of nuclear binding components with 0.4 *M* KCl. A definite peak of ^{3}H-R5020 is found in the 5*S*–6*S* region on the gradients.

As is clear from the data in Table 3, the number of binding sites per DNA for the hypothalamus and hypophysis are almost equal, although the values per mg nuclear protein for the hypophysis are much greater than those for the hypothalamus. This indicates that the same magnitude of progestin binding sites is present in nuclei of both the hypothalamus and anterior hypophysis.

Residual Progestin Binding. It is noteworthy that non-KCl-extractable radioactivity is present in the hypothalamic and hypophyseal nuclear extracts in considerable amounts (Table 3). Residual progestin receptors in brain cell nuclei, extracted with TBDG (Tris-DTT, bacitracin, and glycerol), seemed not to be trans-

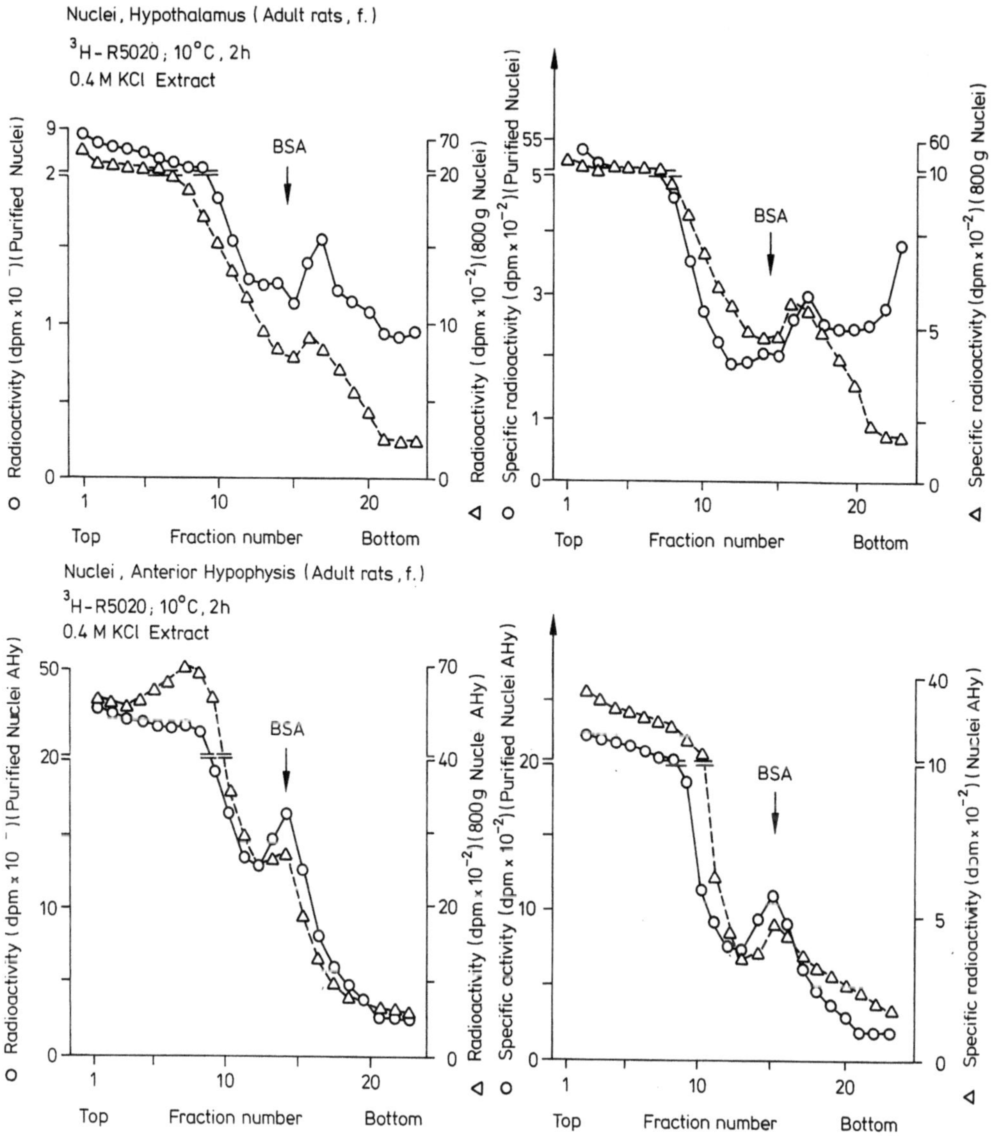

Fig. 3. Sucrose density gradient patterns of the 0.4 *M* KCl-extractable progestin receptors from adult rat hypothalamic (upper panels) and anterior hypophyseal (lower panels) purified and crude nuclei, incubated in vitro with ^{3}H-R5020 at 10 °C. Details of the experimental conditions are described in the text. (○—○) Purified nuclear extract; (△—△) crude nuclear extract. [From Kato and Onouchi (1979)]

Table 3. KCl-extractable and -unextractable progestin binding in purified and crude nuclei from adult female rat hypothalamus and anterior hypophysis.[a] [Adapted from Kato and Onouchi (1979)]

	Hypothalamus		Anterior hypophysis	
	Purified nuclei	800 *g* nuclei	Purified nuclei	800 *g* nuclei
KCl-extractable[b]:				
Fmol/protein mg	93	200	211	312
Fmol/mg DNA	308	837	270	449
KCl-unextractable:				
Fmol/protein mg	350	1,535	442	473
Fmol/mg DNA	1,159	6,432	564	1,000

[a] Hypothalamic purified and crude nuclei were prepared from 1.415 and 1.323 g wet wt., respectively. Purified and crude hypophyseal nuclei were obtained from 0.849 and 0.765 g respectively. Purified and crude (800 g) nuclei were incubated with ^{3}H-R 5020 at 10 °C for 2 h

[b] Radioactivity of the KCl-extractable progestin binding from nuclei incubated with ^{3}H-R 5020 at 10 °C for 2 h. Aliquots (0.05 ml) of the extract were measured directly for radioactivity

located by progesterone. Moreover, estrogen priming increases the concentration of nuclear progestin receptors extractable by 0.4 *M* KCl, but not that of residual receptors (Rainbow et al. 1982a).

In this context, it is interesting to consider the report of Pasqualini et al. (1978) that estrogen receptors resistant to extraction with 0.3–0.4 *M* KCl exist in fetal guinea pig brain. These resistant nuclear sites have also been found for estradiol receptors in immature rat uterus (Clark et al. 1976), and in the kidney, lung, and uterus of fetal guinea pigs (Pasqualini et al. 1976), and for androgen receptors in rabbit epididymis (Danzo and Eller 1976). The function and significance of the residual nuclear sites in the brain are still entirely unknown.

2.1.2.2 Nuclear Exchange at Low Temperatures

Nuclear estrogen exchange in the rat brain has been reported to be effectively completed at higher temperatures (25°–30 °C) (Anderson et al. 1973; Roy and McEwen 1977). In hypothalmic and hypophyseal nuclei, as in those of the uterus (Milgrom et al. 1977; Walters and Clark 1977), binding of ^{3}H-R5020 was achieved by 2 h at 0°–10 °C but rapidly degraded at 30 °C (Kato and Onouchi 1979). Moreover, when unlabeled R5020 was added to the incubation tubes previously incubated with ^{3}H-R5020 at 0°–10 °C, followed by further incubation at 0°, 10°, or 20 °C, the hypothalamic and hypophyseal exchange was found to be completed by 2 h at all temperatures, but the hypothalamic exchange tended to decrease by 4 h at 20 °C (Fig. 4).

2.1.3 Nuclear Receptor Translocation and Cell Receptor Dynamics

Nuclear and cytosolic progestin receptor levels after progesterone injection were studied in the estrogen-primed guinea pig (Blaustein and Feder 1979a, b, 1980)

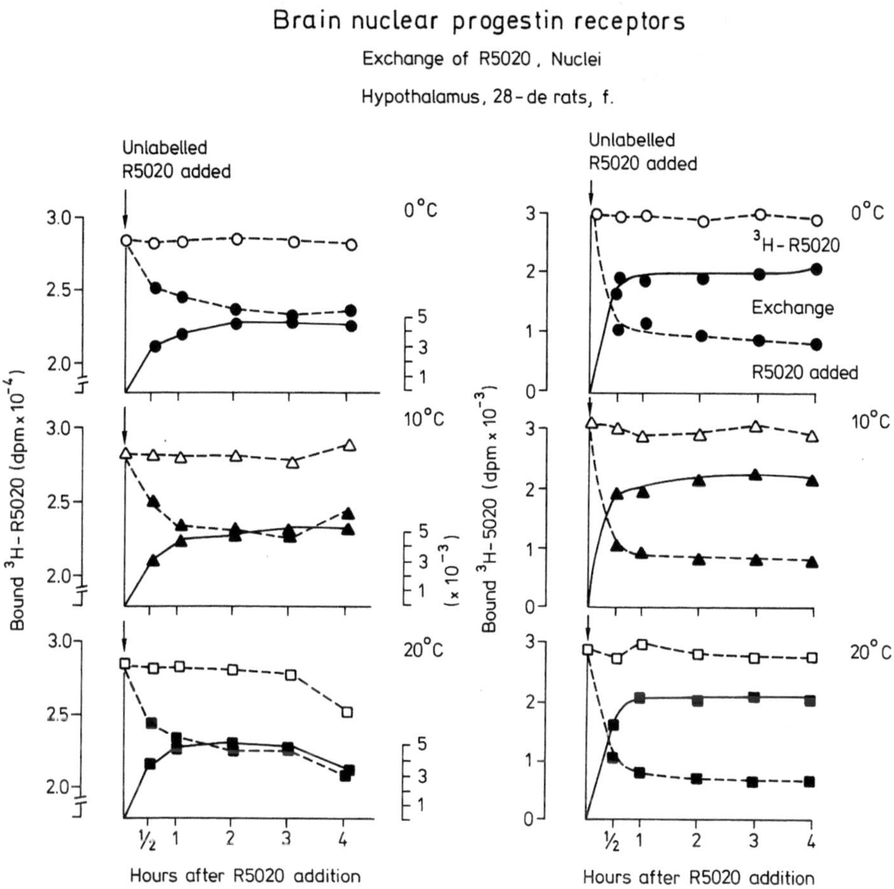

Fig. 4. Progestin exchange in rat hypothalamic and hypophyseal nuclei. Nuclei from hypothalami (56.8 μg DNA) or hypophyses (18.4 μg DNA) of oestradiol-primed 28-day-old rats, treated 1 h in vivo with progesterone, were incubated at 10 °C for 2 h with 12.7 n*M* ^{3}H-R5020. A secondary exchange of unlabeled R5020 for ^{3}H-R5020 bound to the progestin receptors was examined. 1.27 μ*M* unlabeled R5020 was added to tubes previously incubated only with ^{3}H-R5020. The incubation was performed at 0°, 10°, and 20 °C; for 0.5, 1, 2, 3, and 4 h. (---), with ^{3}H-R5020 only (A); (---), unlabeled R5020 added at 0 h (B); (—), A–B. [From Kato and Onouchi (1979)]

and rat (McGinnis et al. 1981 b). As shown in Fig. 5 there is a depletion of cytosolic progestin receptors 2–4 h after progesterone injection, which remains fairly constant until 24 h after progesterone injection, indicating no replenishment of the cytosolic progestin receptors (Blaustein and Feder 1979 b, 1980). Simultaneously with the depletion of cytosolic receptors, there is an accumulation of nuclear progestin receptors, which peaks 2–4 h after progesterone injection. The maximal level in the plasma is seen 2 h after progersterone injection. McGinnis et al. (1981 a) have obtained results on rats similar to those from the guinea pig. Nuclear progestin receptor levels in the hypothalamus were maximal 1 h after intravenous injection of progesterone and declined thereafter, returning to control

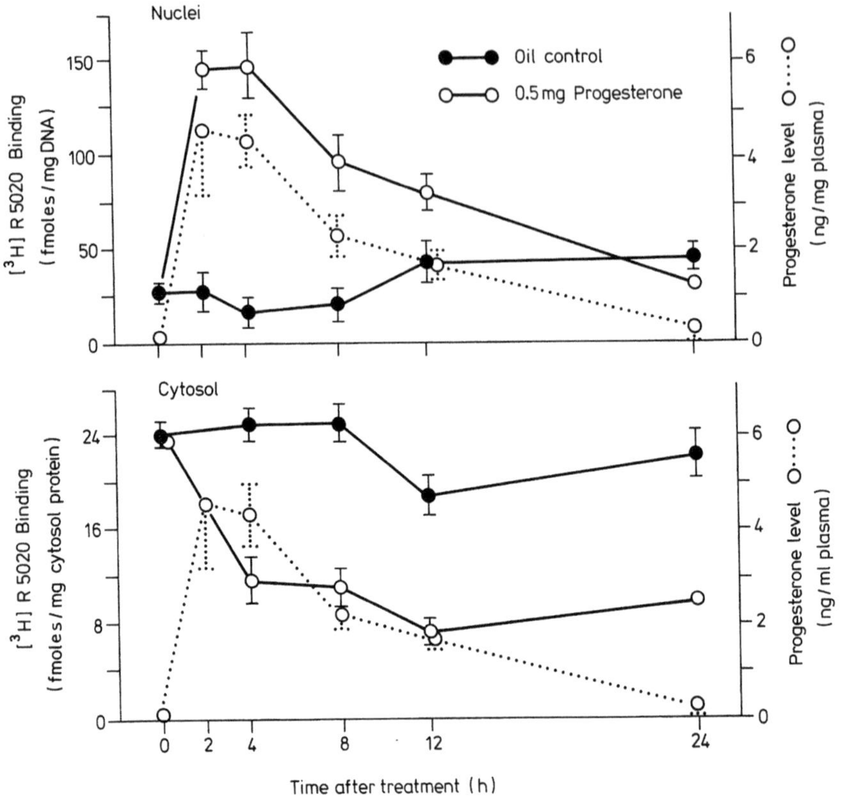

Fig. 5. Time course of depletion of cytoplasmic progestin receptors and accumulation of nuclear progestin receptors in the hypothalamus-preoptic area – septum (HPS), and plasma levels of progesterone, in ovariectomized guinea pigs injected with 1.6 μg estradiol benzoate, followed 40 h later by oil or 0.5 mg progesterone. Animals were killed at various times after oil or progesterone injection, their plasma assayed for progesterone concentration, and the HPS was assayed for either cytosolic or nuclear progestin receptors. Each point is the mean (±SEM) of three to five determinations of receptor content or five to eight independent determinations of the plasma progesterone level. [From Blaustein and Feder (1980)]

levels by 10 h after injection. These results clearly demonstrate that the nuclear translocation of progestin receptors in the hypothalamus correlates with the appearance of feminine sexual behavior and gonadotropin release (see Chap. 4).

Both estrogen-inducible and uninduced progestin receptors in the hypothalamus and cerebral cortex can after progesterone injection translocated into cell nuclei in the same ratio, in adult ovariectomized or ovariectomized-adrenalectomized rats (Rainbow et al. 1982 a) and immature male and female rats at 10 days of age (Kato et al. 1984 b).

In the absence of progesterone, moreover, high doses of estradiol can translocate hypothalamic progestin receptors into cell nuclei (Rainbow et al. 1982 a). This might explain why estradiol can substitute for progesterone in the activation of lordosis behavior (Kow and Pfaff 1977).

2.1.4 Localization of Progestin Receptors in the Rodent Brain

2.1.4.1 Gross Biochemical and Autoradiographic Studies

Two distinct progestin receptor systems exist in the rodent brain: one class of receptors, localized in the hypothalamus and preoptic area, is induced by estrogen; the second class of receptors is unaffected by estrogenic stimulation (MacLusky and McEwen 1978, 1980 a, b; Kato and Onouchi 1979, 1983 a).

Estrogen-inducible progestin receptors have been found biochemically (Moguilewsky and Raynaud 1979 a; Kato and Onouchi 1977, 1979; MacLusky and McEwen 1978, 1980 a; MacLusky et al. 1980; Blaustein and Feder 1979 b) and autoradiographically (Warembourg 1978 b, c) to be mostly concentrated in the mediobasal hypothalamus and preoptic area. Uninduced progestin receptors are present throughout much of the rat brain: in the midbrain, cerebral cortex, amygdala, hippocampus, caudate-putamen, and cerebellum (MacLusky and McEwen 1980 a). In absolute terms, the concentration of "estrogen-insensitive" progestin receptors in brain is relatively low, of the same order as that seen in the non-estrogen-stimulated hypothalamus, preoptic area, and the pituitary. Considerable variation exists in the concentrations of receptors in different brain regions, with the lowest levels in the cerebellum and the highest levels in the cerebral cortex (MacLusky and McEwen 1978, 1980 a). This differential localization of central progestin receptors is essentially similar to that of estrogen receptors, – except in the amygdala (McEwen et al. 1975; Kato 1977 b) – suggesting that induction of the progestin receptors is possibly mediated by estrogen receptors. The inplications of the lack of progestin receptors in the amygdala will be mentioned later (see Sect. 2.3).

2.1.4.2 Quantitative Mapping of Central Progesterone Receptors

Parsons et al. (1982) determined the concentration of cytosolic progestin receptors in 31 nuclei or brain regions removed from frozen sections in rats according to the punch-out method of Palkovits (1973). This was the first quantitative mapping of cytosolic progestin binding in a vertebrate brain.

As shown in Table 4, a basal level (1–8 fmol/mg of protein) of progestin receptors is observed in a variety of preoptic, hypothalamic, and limbic structures in ovariectomized animals. Estrogen treatment induces high levels (24–49 fmol/mg protein) of progestin receptors in regions of the preoptic area (POA) of the hypothalamus which contain high levels of estrogen receptors. Moderate levels (2–8 fmol/mg protein) of progestin receptors are induced by estradiol in other hypothalamic and limbic structures, including the anterior and lateral hypothalamus, the bed nucleus of the stria terminalis, the cingulate cortex, the medial amygdaloid nucleus, and the CA_1 subfield of the hippocampus. In contrast, neither estrogen-inducible nor uninduced progestin receptors are detected in the caudate-putamen and the supraoptic nucleus.

Pathways Connecting Regions Which Contain Progestin Receptors. Anatomical and autoradiographic findings have indicated that estrogen-concentrating regions in the rodent brain are linked by neural pathways (Pfaff and Kleiner 1973; McEwen and Pfaff 1973). Such fiber pathways could provide an anatomical sub-

Table 4. Mapping of cytosolic receptors for progestin and estrogen in brain nuclei and subregions of the female rat brain. [Adapted from Parsons et al. (1982); Rainbow et al. (1982c)]

Region	Progestin receptor (fmol/mg protein)[a]		N	Estrogen receptor (fmol/mg protein)	N
	$-E_2$	$+E_2$			
Diagonal band of Broca	2.1±1.4	1.9±1.1	4	2.8±0.7	4
Medial septum	2.7±1.6	0.9±0.7	4	N.S.	6
Lateral septum	1.9±1.7	1.0±0.6	5	1.8±0.8	6
Olfactory tubercle	1.4±0.5	1.7±0.7	4	2.2±0.4	4
Nucleus stria terminalis	1.4±0.9	4.6±1.2	5	6.1±1.3	6
Medial preoptic nucleus	6.2±2.2	41.8±3.5	6	21.6±4.9	6
Suprachiasmatic preoptic nucleus	5.1±3.3	35.8±7.8	5	13.2±2.9	4
Periventricular preoptic nucleus	1.9±1.5	48.8±8.1	5	32.0±1.0	6
Lateral preoptic nucleus	1.4±0.7	4.9±1.5	6	1.9±0.8	5
Caudate-putamen	0.5±0.4	1.3±1.2	5	N.S.	4
Parietal cortex	1.7±0.5	2.3±0.8	5	N.S.	6
Cingulate cortex	2.0±1.0	4.9±1.2	5	1.4±0.2	4
Periventricular anterior hypothalamus	7.4±3.3	28.9±4.9	6	18.5±1.3	6
Anterior hypothalamus	2.2±0.8	7.5±2.1	4	6.7±2.5	5
Supraoptic nucleus	3.9±3.9	3.3±3.3	4	8.9±1.7	6
Ventromedial nucleus	4.5±1.1	24.7±4.0	6	12.0±1.7	6
Lateral ventromedial nucleus	8.9±3.4	24.4±3.2	4	–	–
Medial ventromedial nucleus	3.6±3.6	12.7±2.1	3	–	–
Dorsomedial nucleus	2.8±1.1	5.3±0.9	5	4.9±0.7	6
Lateral hypothalamus	1.4±0.7	4.9±1.5	6	14.2±1.0	4
Arcuate-median eminence	6.3±1.0	40.9±5.3	6	13.4±1.7	6
Medial amygdala	1.3±0.7	2.8±1.0	5	9.8±1.7	6
Cortical amygdala	0.4±0.2	3.5±1.5	6	3.4±0.5	6
Ventral thalamus	3.2±1.7	2.3±1.3	4	N.S.	5
Dentate gyrus	1.7±1.9	0±0	5	3.0±1.3	5
CA_1	1.0±0.6	5.1±1.1	5	N.S.	5
CA_3	0.8±0.4	0.4±0.4	5	N.S.	5
Ventral premammillary nucleus	2.4±1.5	3.3±2.5	4	5.4±1.5	4
Habenula	1.5±0.9	3.0±1.8	4	1.3±0.6	4
Central grey	2.2±1.0	3.8±2.0	5	N.S.	4

N, Number of experiments; N.S., not significantly different from a zero level of receptors
[a] The values are expressed as femtomoles of ^{3}H-steroid per mg of protein±SEM. ^{3}H-R 5020 was used to determine for progestin receptor levels and ^{3}H-estradiol for estrogen receptors

strate for coordinated and systematic physiological changes by estradiol. Progesterone receptor-containing nuclei and brain regions are also interconnected by neural pathways. Such pathways may provide an anatomical substrate for integrated actions of progesterone and estrogen in the female rat brain (Parsons et al. 1982).

2.1.4.3 Summary

Gross biochemical, autoradiographic and mapping studies have revealed that in the rodent brain, estrogen-inducible progestin receptors are distributed similarly to the estrogen receptors, through which estrogen may act to induce progestin receptors in the brain and hypophysis. Moreover, as neural pathways connect estrogen-concentrating regions, progestin receptor-containing nuclei and brain regions could also be connected by neural pathways. It is particularly of interest that the amygdala contains few if any progestin receptors even in estrogen-primed rats (MacLusky and McEwen 1978, 1980 a; Kato and Onouchi 1979), despite relatively high levels of estrogen receptors in the region (Pfaff and Kleiner 1973; Stumpf and Sar 1969; Stumpf 1970; Warembourg 1970; Maurer and Wooley 1974).

Microassay data on estrogen and progestin receptors in the rat brain (Rainbow et al. 1982 c; Parsons et al. 1982) have confirmed that some estrogen receptor-containing regions, such as the cortical amygdala, do not exhibit induction of progestin receptors by estradiol. This may be the case with the monkey amygdala (MacLusky et al. 1980; Kato et al. 1981, see Sect. 2.3.3.) There is evidence for a direct action of progesterone on the amygdala (Piva et al. 1973). The implication is that progesterone's effect is not receptor-mediated in this specific region. This aspect will be discussed in Sect. 2.3.

2.2 Birds

In birds, progesterone plays a role in the regulation of the reproductive function through its action on the hypothalamus and/or the central nervous system (Ralph and Fraps 1960; Komisaruk 1967; Furr et al. 1973). Despite the well-established data on progesterone receptors in peripheral targets such as chick oviduct (O'Malley and Means 1974), there are few reports on progesterone receptors in the central targets of the hen.

Kawashima et al. (1978) demonstrated the existence of specific ^{3}H-progesterone binding sites with a high affinity and limited capacity in the cytosols of the hypothalamus and hypophysis from the laying hen (*Gallus domesticus*). Nuclear progesterone receptors were also demonstrated, by ^{3}H-progesterone exchange, in the hen hypothalamus and hypophysis. The values of K_d for the tissues were in order of $10^{-10} M$. The number of binding sites was 21–29 fmol/100 µg DNA. The existence of progesterone receptors in cytosols and nuclei from the hypothalamus and hypophysis suggests a direct action of the hormone on the hypothalamus and hypophysis in the hen (Kawashima et al. 1979). Balthazart et al. (1980) have also demonstrated the existence of progestin receptors in the brain cytosols of male ring doves (*Streptopelia risoria*). Its concentration was higher in the anterior hypothalamus-preoptic area than in other brain regions (the posterior hypothalamus, hyperstriatum, and midbrain). It is interesting that the whole hypothalamus, even in castrated birds, had a concentration of binding sites (60 fmol/mg protein or more) much higher than those (on the order of 10 fmol/mg protein) in the rat and guinea pig. Steroid specificity for avian brain progestin receptors seems to be somewhat less rigid than in the rodent brain; corticosterone competes well for binding sites in the dove and hen, but not in the rat (Kato and Onouchi 1977; MacLusky and McEwen 1978) or guinea pig (Blaustein and Feder 1979 a).

Hormones modulate the concentration of cytosolic progestin receptors in the male dove brain (Balthazart et al. 1980). Testosterone priming increases the concentration of receptors in the hypothalamus, but not in the hyperstriatum and midbrain. The sensitivity to progesterone of the dove brain may change as a consequence of increases in the level of testosterone in the circulation. This effect of testosterone could be mediated by the aromatization of testosterone in brain tissue.

2.3 Studies on the Monkey Brain

Although knowledge about steroid-receptor interactions in the central nervous system is essential for understanding the neural control of reproduction in primates, available information is still limited, especially on progestin binding and receptors in the primate brain (MacLusky et al. 1980; Garris et al. 1981; Kato et al. 1981).

2.3.1 Rhesus Monkey

Garris et al. (1981) have reported in situ labeled estradiol and progestin (R5020) localization in the vascularly separated and isolated hypothalamus of the rhesus monkey. Autoradiographically, the infundibular nucleus (i.e., the arcuate nucleus) and caudal ventromedial hypothalamic (VMH) regions are densely labeled. Tritiated R5020 label is observed in the cells of the preoptic-rostral hypothalamic areas, the organum vasculosum of the lamina terminalis (OVLT), and the areas between the infundibular and medial mammillary nuclei. The localization of progestin-concentrating cells in the rhesus monkey is similar to that in the female rat (Warembourg 1978 b, c) and female guinea pig (Sar and Stumpf 1973; Warembourg 1978 a). The localization of ^{3}H-estradiol-concentrating neurons within the preoptic, anterior, and medial hypothalamic regions in the rhesus monkey following systematic administration of ^{3}H-estradiol (Garris et al. 1981) confirms the findings of Pfaff et al. (1976). Tritiated ^{3}H-estradiol appears to disperse throughout the hypothalamus.

These findings are consistent with the notion that the center of gonadotropin control in the female rhesus monkey resides within the mediobasal hypothalamus-pituitary axis (Knobil 1974; Krey et al. 1975; Spies et al. 1977; Hess et al. 1977; Ferin et al. 1979).

2.3.2 Bonnet Monkey

MacLusky et al. (1980) have demonstrated putative progestin receptors, using gel filtration on LH-20 and sucrose density gradient centrifugation, in soluble fractions of the anterior hypophysis and some brain areas of the Bonnet monkey (*Macaca radiata*). Progestin-specific components with high affinity ($K_d \simeq 0.3\text{n}M$) sediment in the $6S$–$7S$ region on 10% glycerol-sucrose gradients. In ovariectomized monkeys, the progestin receptors are restricted to the hypothalamus and preoptic area. No receptors are detected in the cingulate or parietal cortex, amygdala, midbrain, or cerebellum. Estrogen priming, via implantation of estradiol-containing

Silastic capsules to generate serum levels of estradiol in the physiological range, causes marked elevation of progestin receptor levels in the anterior hypophysis (10-fold) and in the hypothalamus (3-fold), but not in any other brain area studied.

Although the physicochemical character of cytosolic progestin receptors seems to be similar in the rat and Bonnet monkey, there are two major interspecies differences in receptor localization and induction (MacLusky et al. 1980). While the monkey brain progestin receptors are restricted to the hypthalamus and preoptic areas, the rat brain receptors are more widely distributed. A second major interspecies difference concerns the ability of estrogen treatment to induce cytsolic progestin receptors in the preoptic area. Estrogen-priming can increase cytosolic progestin receptor levels in the preoptic area of the rat, but not in the Bonnet monkey. It has been suggested that these differences reflect a divergence between rodents and primates in the sites of ovarian steroid hormone feedback on brain and pituitary function.

2.3.3 Studies on the Japanese Monkey

Our preliminary report demonstrated progestin and estradiol receptors in the brain and hypophysis of the estrogen-primed Japanese monkey (*Macaca fuscata*) (Kato et al. 1981). More detailed studies have now been carried out.

2.3.3.1 Materials and Methods

Monkey Treatment and Brain Dissection. Five adult female Japanese monkeys (*Macaca fuscata fuscata*), weighing 6.3–9.0 kg, were used. The monkeys were fed once a day with monkey chow, and twice a week with vegetables and fruit. The animals were bilaterally ovariectomized under Ketamine HCl (Ketalar, Sankyo, Tokyo) anesthesia (20 mg/kg, i.m.), and an antibiotic (Keflex, Shionogi, 150 mg/day) was given once a day for 3 consecutive days. Starting at least 1 week after castration, the monkeys were injected s.c. once daily with 200–400 μg 17β-estradiol benzoate (Pellanin, Mochida Seiyaku, Tokyo) in sesame oil, for 4–14 consecutive days. The animals were bled completely through the jugular artery under Ketamine HCl anesthesia (20 mg/kg body weight, i.m.) 24 h after the last injection of estradiol benzoate: the brain and hypophysis were removed, rinsed in ice-cold physiological saline and placed on an ice-cold glass plate.

Each brain was dissected according to the map of Kusama and Mabuchi (1970) into the preoptic area-anterior hypothalamus, posterior hypothalamus, A23 area, amygdala, hippocampus, midbrain, parietal and cingulate cortexes, cerebellum, callosum, and thalamus. In some, the fasciculus opticus, corpus quadrigemina, area postrema, pons, and spinal cord were also dissected. The tissues were rinsed in ice-cold physiological saline, blotted on filter paper, immediately frozen, and kept at −80 °C until used.

Preparation of Nuclear and Cytosolic Fractions. Tissues were homogenized in ice-cold TETG buffer [10 m*M*Tris-HCl, 1 m*M* sodium EDTA, 12 m*M* thioglycerol and 10% (vol/vol) glycerol, pH 7.4] using a Teflon homogenizer (Takashima, Tokyo); five strokes, 600–1,000 rpm, 3 min) in ice. Homogenization volumes (vol ml/wet weight gr) were as follows: brain and neural tissues, 1.4–1.5; and hypophyses, 5.0–6.0. The homogenates were then centrifuged at 800 *g* for 20 min to obtain a nuclear fraction and low-speed supernatant fraction. The nuclear fraction was immediately suspended in 0.25 *M* sucrose TMKG buffer (0.25 *M* sucrose, 10 m*M* Tris-HCl, 12 m*M* thioglycerol, 2 m*M* $MgCl_2$, 10 m*M* KCl and 10% glycerol, pH 7.5) and centrifuged at 800 *g* for 20 min. The pellet was resuspended and sedimented three more times; once in 0.25 *M* sucrose-TMKG buffer containing 0.5% Triton X-100 and twice in 0.25 *M* sucrose-TMKG buffer. The final pellet was suspended in 0.25 *M* sucrose-TMKG buffer and was used for the measurement of nuclear estrogen receptors. The low-speed supernatant fraction was spun at 105,000 *g* for 1 h (Hitachi 65P model ultracentrifuge, Hitachi Co., Tokyo) to yield the high-speed supernatant (cytosol). The cytosol fraction was then treated with dextran-coated charcoal to remove free hormone. All subsequent procedures, unless otherwise noted, were performed at 4 °C.

Sucrose Density Gradient Centrifugation. The supernatant fraction cytosol (0.29 ml) was incubated with gentle shaking for 4 h at 0 °C with ^{3}H-R5020 [specific activity (SA) 87.1 Ci/mmol] in the absence or presence of a 100-fold excess of unlabeled steroid hormones dissolved in 5 µl ethanol.

The mixture, layered on a sucrose-10% glycerol density gradient, was centrifuged in a Beckman model L3–50 or a Hitachi RP65 ultracentrifuge as previously described (Kato and Onouchi 1977). After density gradient centrifugation, the tubes were fractionated (0.2-ml fractions) and counting was performed as previously described (Kato and Onouchi 1977). To determine apparent sedimentation coefficients (*S*) by the method of Martin and Ames (1961), crystalline bovine albumin (4.6 *S*) and yeast alcohol dehydrogenase (7.6*S*) were used as standards.

Measurement of Binding by Dextran-Coated Charcoal Absorption. To obtain data for the analysis of binding kinetics and Scatchard plot (Scatchard 1949), the cytosol fraction (0.3 ml; 150 µg protein) was incubated for 16 h at 4 °C with various concentrations of ^{3}H-labeled steroid (0.1–2.0 n*M*), in the absence or presence of a 100-fold excess of unlabeled hormone. To absorb unbound radioactivity, 1 ml of a suspension of dextran-coated charcoal (0.25% Norit A and 0.0025% Dextran T-70 in TETG buffer, pH 8.0) was added to each tube, mixed, and incubated for 10 min at 4 °C, followed by centrifugation for 10 min at 1,500 *g* in a refrigerated centrifuge. A 1-ml aliquot of the supernatant was transferred to a scintillation vial for counting with 10 ml scintillation fluid consisting of 4 g Omnifluor (NEN, Boston, MA) dissolved in 666 ml toluene and 333 ml Triton X-100. The radioactivity was measured in an Aloka model LSC-653 scintillation counter (Aloka Co., Tokyo) and a Nuclear-Chicago Co. Mark II liquid scintillation spectrometer.

Specific binding was obtained by subtracting nonspecific binding (^{3}H-steroid plus a 100-fold excess of unlabeled hormone) from total binding (^{3}H-steroid alone). The number of binding sites (NBS) and dissociation constants (K_d) were determined by means of the Scatchard plot.

Determination of Nuclear Estrogen Receptors by Exchange. The estrogen receptor content in nuclear fractions was determined using a modification of the ^{3}H-estradiol nuclear exchange technique described by Anderson et al. (1972). Suspended nuclei, in 0.3 ml 0.25 M sucrose-TMKG buffer containing 8.6 nM ^{3}H-estradiol (for total exchange), or 8.6 nM ^{3}H-estradiol plus 0.86 μM diethylstilbestrol (for nonspecific exchange), were incubated at 30 °C, with shaking. After cooling to 4 °C, the nuclear fractions were washed as previously described. The ^{3}H-estradiol was then extracted from the nuclear pellet with 1.5 ml absolute ethanol, and a 1-ml aliquot of extract was added to 10 ml scintillation fluid (4 g Omnifluor in 1 liter toluene). Radioactivity was determined as previously described.

Miscellaneous Assays. Protein was determined by the method of Lowry et al. (1951), using bovine serum albumin as the standard. DNA was quantitated by the method of Burton (1956), using calf thymus DNA as the standard.

Steroids and Chemicals. 17β-[1,2,6,7-^{3}H]Estradiol (SA 91.8 Ci/mmol), [17α-methyl-^{3}H]17α-,21-dimethyl-9-nor-4,9-pregnadiene-3,20-dione (^{3}H-R5020; SA 87.07 Ci/mmol) and radioinert R5020 were obtained from New England Nuclear Co. (Boston, MA). The purity of these radioactive compounds was periodically checked by thin layer chromatography as previously described (Kato and Villee 1967 a).

17α-Estradiol-3-benzoate (Pellanin) was supplied by Mochida Seiyaku (Tokyo). Other steroids and diethylstilbestrol (DES) were purchased either from Sigma Chemical Co. (St. Louis, MO) or Steraloids Inc. (Pawling, NY). All other chemicals were reagent or analytical grade.

2.3.3.2 Results

Identification of Cytosolic Progestin Receptors in the Japanese Monkey Brain and Hypophysis. A definite peak of ^{3}H-R5020 binding is found in the vicinity of 7*S* in hypophyseal gradients from the estrogen-primed monkey (Fig. 6). Multiconcentration saturation analysis of hypophyseal ^{3}H-R5020 binding revealed linear high-affinity–low-capacity progestin binding. Figure 7 shows representative linear Scatchard plots of ^{3}H-R5020 binding in the preoptic area-anterior hypothalamus (HPOA), the posterior hypothalamus, and the A23 region located just rostral to the preoptic region. Dissociation constants (K_d) of progestin binding in the brain and hypophysis were on the order of 0.1–0.3 nM. The binding in these brain tissues was found to be high affinity and low capacity. These results clearly indicate the presence of specific progestin receptors in the Japanese monkey brain and anterior hypophysis.

Distribution of Cytosolic Progestin Receptors and Nuclear Estradiol Receptors in the Brain. The concentration of cytosolic progestin receptors and nuclear estradiol receptors were simultaneously determined in various parts of the brain and hypophysis of the estrogen-treated Japanese monkey. One example of the distribution of both receptors in the monkey (Mff-2) brain and hypophysis is shown in Fig. 8.

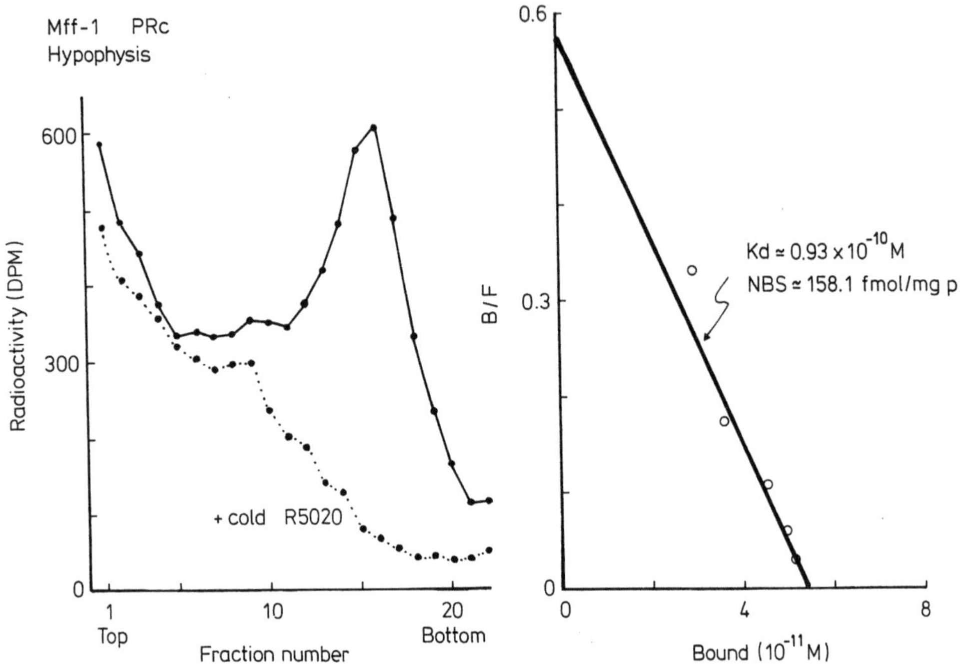

Fig. 6. Representative sucrose density gradient pattern and Scatchard (1949) plot of progestin binding in anterior hypophyseal cytosol from an estrogen-primed Japanese monkey (*Macaca fuscata fuscata,* Mff-1). The cytosol was incubated at 0 °C for 4 h with ^{3}H-R5020, (—); or with ^{3}H-R5020 and unlabeled R5020 (····)

The data for cytosolic progestin receptors and nuclear estradiol receptors in the central neural structures and hypophysis, based upon data obtained from the five monkeys, are compared in Fig. 9. High concentrations of cytosolic progestin receptors are found in the anterior hypophysis and the HPOA. The posterior hypothalamus and the A23 area concentrate ^{3}H-R5020 considerably. Little or no progestin binding is associated with the cerebral cortex, cerebellum, hippocampus, amygdala, reticular formation, and other brain regions and neural structures examined. High levels of nuclear estradiol receptors are detected in the anterior hypophysis and HPOA. The posterior hypothalamus and the amygdala concentrate estrogen receptors in their cell nuclei. Considerable but variable estrogen receptor levels are also detected in some neural structures including the reticular formation, corpus quadrigemina, hippocampus, and callosum. Low levels are found in the cerebral cortex, but little or no binding is found in the cerebellum, thalamus, area postrema, pons, or spinal cord. The results from the Japanese monkey brain are essentially similar to those from the rhesus monkey (Gerlach et al. 1976).

When the localization of cytosolic progestin receptors and nuclear estrogen receptors are compared in the same monkey brain, it is noticeable that both receptors are mostly concentrated in the anterior hypophysis, HPOA, posterior hypothalamus, and A23 area. This suggests that estrogen induces cytosolic proges-

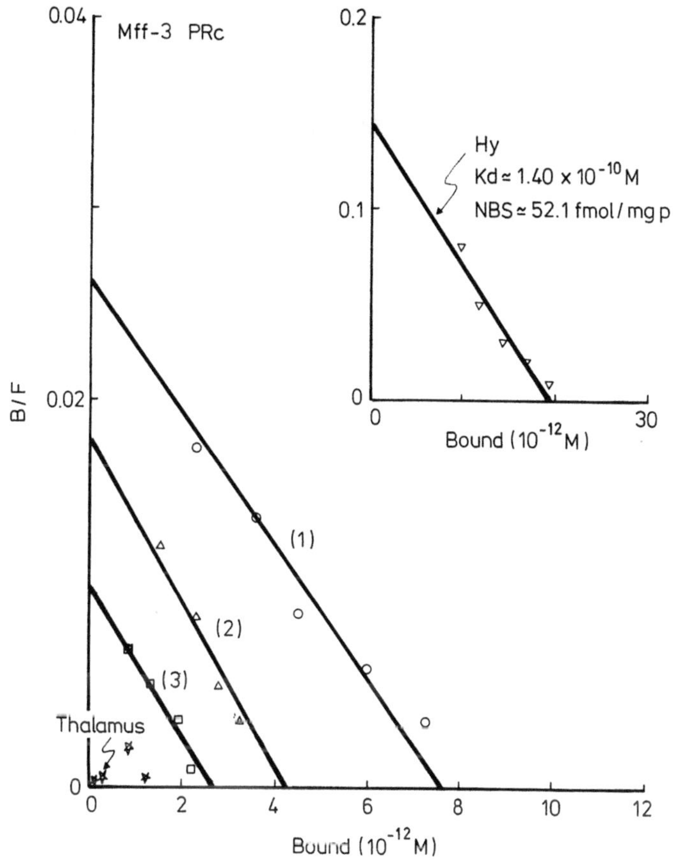

Fig. 7. Saturation analysis of progestin binding in cytosols (PRc) from the preoptic-anterior hypothalamus (PO-Ant. Ht), posterior hypothalamus (Post. Ht), anterior hypophysis (Hy), A23 region, and thalamus from an estrogen-primed Japanese monkey (*Macaca fuscata fuscata*, Mff-3). The cytosols were incubated at 0 °C for 4 h in the presence of varying concentrations of [^{3}H] R5020 or [^{3}H] R5020 with a 100-fold excess of R5020. The amounts of specifically bound [^{3}H] 5020, were measured by dextran-coated charcoal assay, and plotted by the method of Scatchard (1949). (*1*) PO-Ant. Ht, Kd$\simeq 2.96 \times 10^{-10}$ M, NBS$\simeq$16.8 fmol/mg p; (*2*) Post. Ht, Kd$\simeq 1.84 \times 10^{-10}$ M, NBS$\simeq$8.3 fmol/mg p; (*3*) A23 area, Kd$\simeq 2.57 \times 10^{-10}$ M, NBS$\simeq$6.0 fmol/mg p

tin receptors in the monkey brain and hypophysis. Special attention should be paid to the presence of nuclear estrogen receptors and lack of cytosolic progestin receptors in the amygdala of the estrogen-primed Japanese monkey.

2.3.3.3 Discussion and Summary

Cytosolic Progestin Receptor Induction by Estrogen

The almost identical distribution of estrogen and progestin receptors in some specific brain regions and in the hypophysis, together with the findings of MacLusky et al (1980), indicate the induction of cytosolic progestin receptors by estrogen in

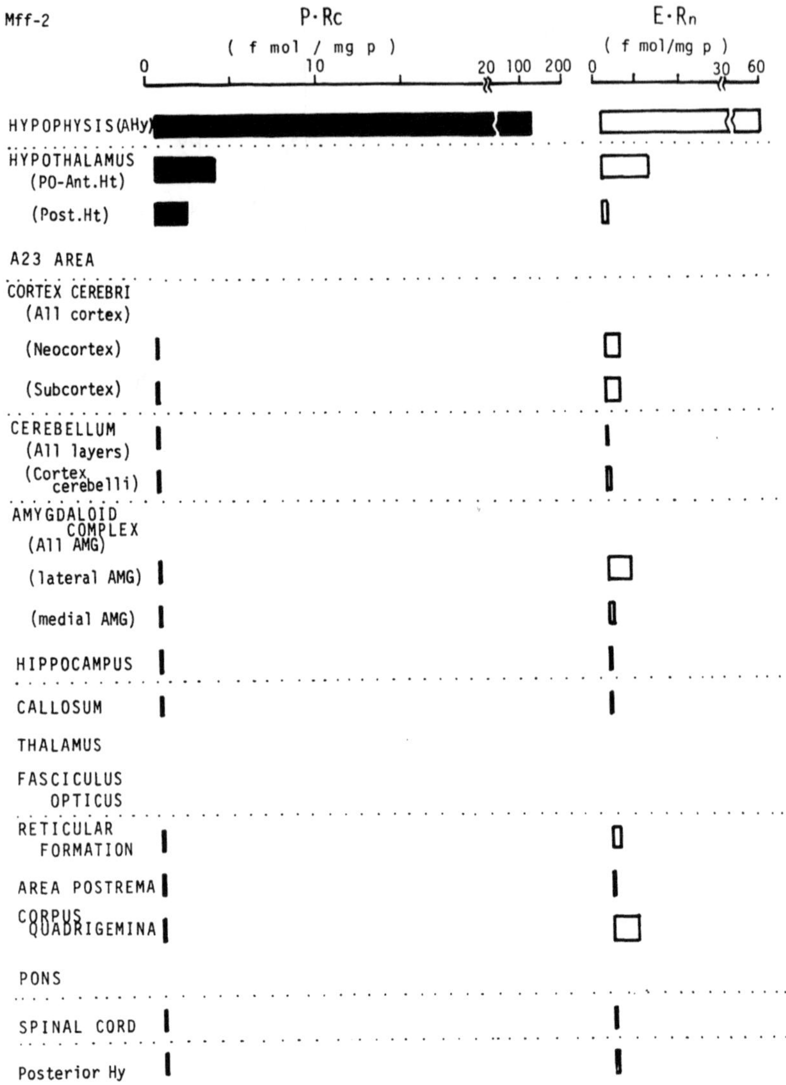

Fig. 8. Representative distribution of cytosolic progestin receptors (P · Rc) and nuclear estradiol receptors (E · Rn) in various parts of the brain and hypophysis from the estrogen-primed Japanese monkey (*Macaca fuscata fuscata*, Mff-2). Determination of cytosolic progestin receptors was carried out by multiconcentration saturation analysis of ^{3}H-R5020 binding. Nuclear estradiol receptors were determined by nuclear exchange assay. Details of the procedure are described in the text

Fig. 9. Summary of the distribution of cytosolic progestin receptors and nuclear estradiol receptors in various parts of the central nervous system and hypophysis of 5 estradiol-primed Japanese monkeys (*Macaca fuscata fuscata*). Cytosolic progestin receptors were determined by multiconcentration analysis of ^{3}H-R5020 binding. Nuclear estradiol receptors were determined by one-point exchange assay. It is notable that the hypothalamus, the preoptic area, and the A23 region just rostral to the preoptic area concentrate both receptors, but progestin receptors are deficient in the amygdala, although it does contain estradiol receptors. PRc, Cytosolic progestin receptors; ERn, nuclear estradiol receptors; N, number of determinations

Fig. 9.

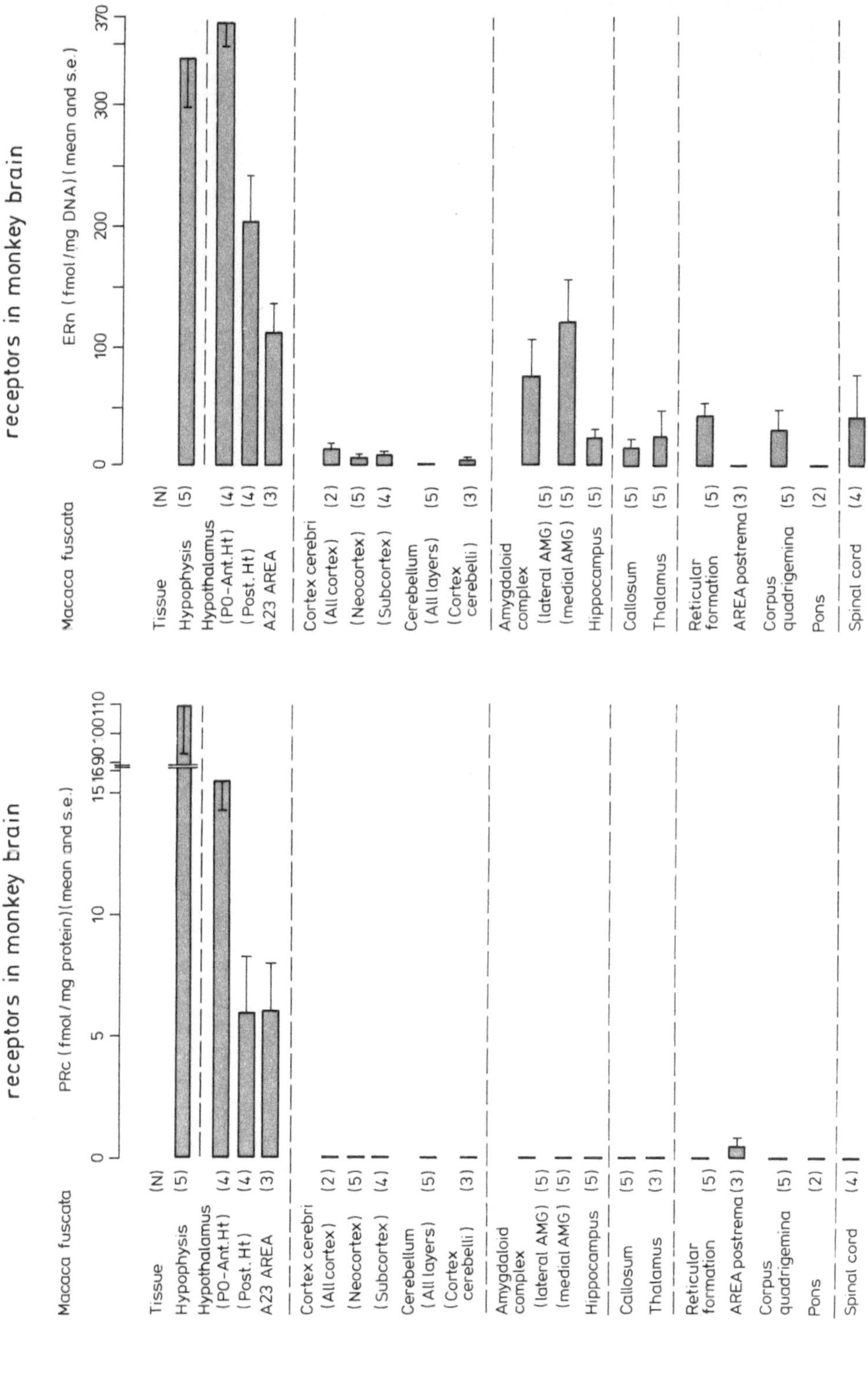
Distribution of cytoplasmic progestin receptors in monkey brain
Macaca fuscata
PRc (fmol / mg protein) (mean and s.e.)
0
5
10
15
1690
100110
Distribution of nuclear estrogen receptors in monkey brain
ERn (fmol / mg DNA) (mean and s.e.)
0
100
200
300
370
Tissue (N)
Hypophysis (5)
Hypothalamus
(PO-Ant.Ht) (4)
(Post. Ht) (4)
A23 AREA (3)
Cortex cerebri
(All cortex) (2)
(Neocortex) (5)
(Subcortex) (4)
Cerebellum
(All layers) (5)
(Cortex cerebelli) (3)
Amygdaloid complex
(lateral AMG) (5)
(medial AMG) (5)
Hippocampus (5)
Callosum (5)
Thalamus (3)
Thalamus (5)
Reticular formation (5)
AREA postrema (3)
Corpus quadrigemina (5)
Pons (2)
Spinal cord (4)

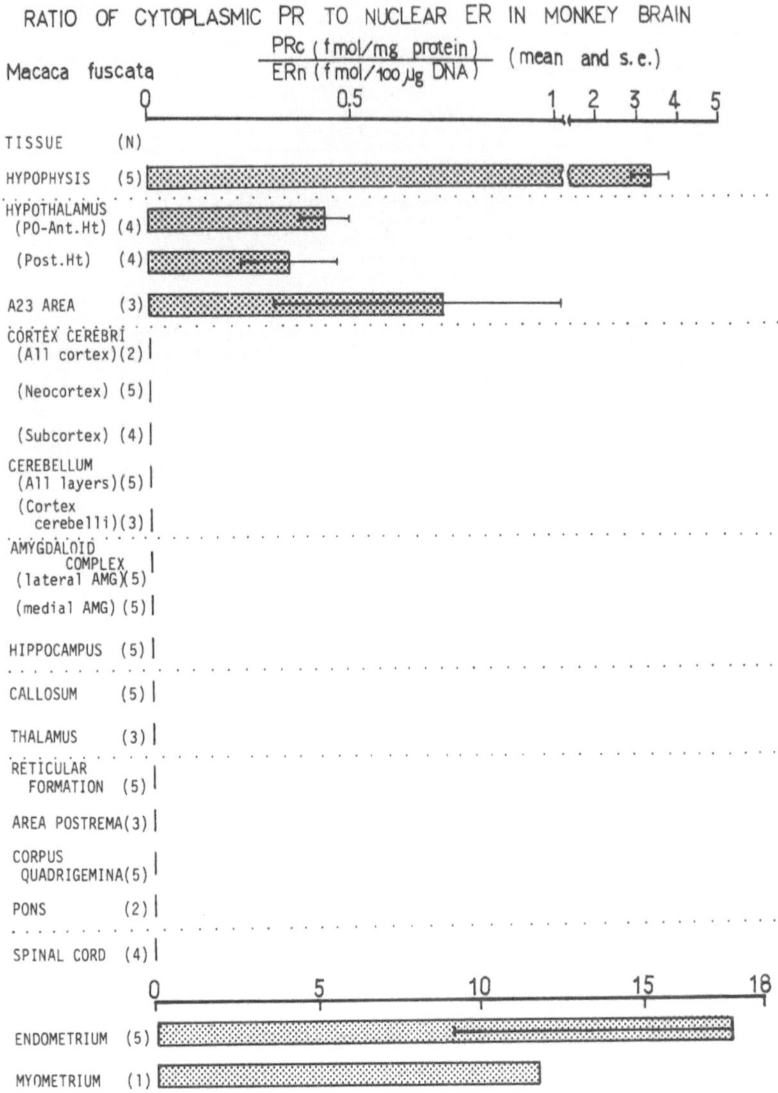

Fig. 10. Ratios of cytosolic progestin receptors (PR) to nuclear estradiol receptors (ER) in estradiol-primed Japanese monkeys (*Macaca fuscata fuscata*)

these regions. Within the Japanese monkey brain, the HPOA has been found to contain the highest levels of progestin and estrogen receptors. Moreover, considerable numbers of both receptors are concentrated in the A23 area, which is located just to the preoptic area, when the brain is dissected according to the map of Japanese monkey brain (Kusama and Mabuchi 1970). The ratios of cytosolic progestin receptors (fmol/mg protein) to nuclear estrogen receptors (fmol/100 µg DNA) were computed for the different regions of the neural structures; the mean values for the HPOA, posterior hypothalamus, and A23 area are 0.42, 0.29, and

0.56, respectively (Fig. 10). The values for other brain regions are zero. This suggests that cytosolic progestin receptors in the HPOA, posterior hypothalamus, and A23 area may be estrogen-inducible. The localization of cytolosic progestin receptors in the A23 area and HPOA region of the Japanese monkey brain is consistent with the findings of Garris et al. (1981) on the rhesus monkey brain, in which the cells of the preoptic-rostral hypothalamic area and the OVLT retain ^{3}H-R5020 label. However, in the Bonnet monkey, the preoptic progestin receptors seem not to be estrogen-inducible (MacLusky et al. 1980). The interspecies difference may be a possible explanation.

Few data are available on the time course of the estrogen-dependent appearance of progestin receptors in the primate brain. For this purpose the monkey sexual skin, a hormone responsive tissue, can be used as a parallel target because of the presence of sexual skin progestin receptors, and because it is possible to take consecutive samples of tissue from the same monkey without killing it. Progestin receptors in the sexual skin have been shown to increase rapidly after estrogen-priming (Kato et al. 1980; Onouchi and Kato 1983). Together with the rodent brain data (McEwen 1978 b), these results suggest that induction of estrogen-dependent progestin receptors may occur rapidly in the monkey brain.

Noninducibility of Amygdala Progestin Receptors by Estrogen. The high levels of progestin receptors corresponded well to the high levels of estrogen receptors in the hypothalamus and anterior hypophysis of the Japanese monkey, and of the bonnet and rhesus monkeys (MacLusky et al. 1980; McEwen 1978 b). These studies revealed little if any progestin binding in the monkey amygdala, despite the relatively high accumulation of estrogen receptors. This implies that estrogen cannot induce progestin receptors in the monkey amygdala.

The concentration of cytosolic progestin receptors is also low in the amygdala of the estrogen-primed rat. There is a possibility that the most of the cytolosic progestin receptors are translocated to the nuclear fraction. To negate this possibility, we have measured nuclear progestin receptor complexes in the amygdala. The estrogen-primed rats were injected with 2 mg progesterone 1 h before killing. Low-temperature exchange of ^{3}H-R5020 for bound progestin, and extraction with potassium chloride, were carried out with nuclei from the amygdala, hypothalamus, and anterior hypophysis (Kato et al. 1978; Kato and Onouchi 1979). The hypothalamic and hypophyseal extract was found to contain a single peak of ^{3}H-R5020 exchanged for progestin, in the region of 4*S*. No definite peak was recognized in the amygdala extract. Thus, the amygdala seems to contain only a low level of progestin receptors, although estrogen receptors are present in the cytosol and nuclei from the amygdala.

Possible Differential Central Actions of Progesterone. The hypophysis and different brain areas of the monkey (and, possibly, rodent) have been categorized in terms of estrogen receptors (ER) and progestin receptors (PR) into ER+PR+, ER+PR−, and ER−PR− groups (Table 5). Progesterone may act differentially on the brain. Progestin receptors, possibly estrogen-inducible, are concentrated mostly in the HPOA, posterior hypothalamus, and A23 area; as well as in the anterior hypophysis. This suggests a central action of progesterone

Table 5. Estrogen and progestin receptors in the brain and hypophysis of *Macaca fuscata fuscata*

Receptors	Region
ER+, PR+	Median eminence
	Preoptic-anterior hypothalamus
	Posterior hypothalamus
	A23 area
	Anterior hypophysis
ER+, PR−	Amygdala
	(Hippocampus)
ER−, PR−	Cerebral cortex
	Cerebellum

Abbreviations: ER, estrogen receptors; PR, progesterone receptors

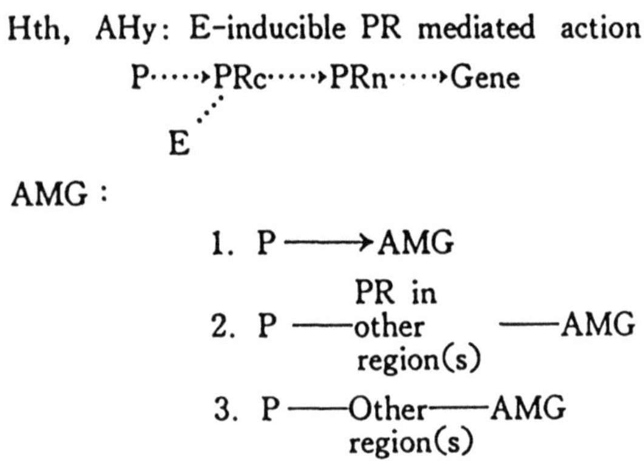

Fig. 11. Differential action of progesterone on the brain and hypophysis. Hth, Hypothalamus; AHy, anterior hypophysis; E, estrogen; P, progesterone; PR, progesterone receptor; PRc, cytosolic progesterone receptor; PRn, nuclear progesterone receptor; AMG, amygdala

through the receptors in the specific brain regions. The absence or very low level of progestin receptors in the amygdala suggests that progesterone does not act directly on the amygdala, especially in the monkey. Further consideration might be given, however, to a direct but non-receptor-mediated action or an indirect effect of progesterone on the amygdala (Fig. 11).

3 Steroidal Modulation of Brain Progestin Receptors

3.1 Induction of Progestin Receptors by Estrogen

Estradiol markedly increases the levels of progestin receptors in the hypothalamus and anterior hypophysis in intact female immature and ovariectomized adult rats (Kato and Onouchi 1977; Kato et al. 1978; Moguilewsky and Raynaud 1977,

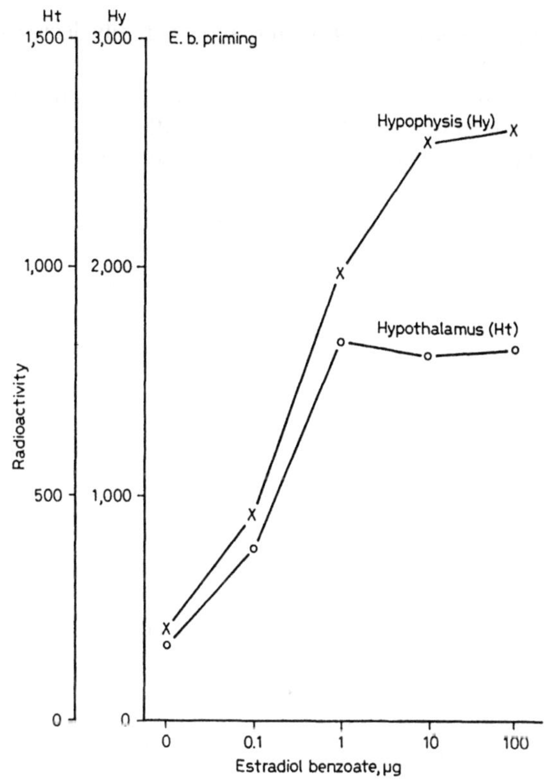

Fig. 12. Dose-dependent induction by estradiol of hypothalamic and anterior hypophyseal progesterone receptors in immature female rats (27-day-old). Estradiol benzoate (E. b.) was injected s. c. once daily for 5 consecutive days prior to sacrifice. [From Kato (1978)]

1979a; MacLusky and McEwen 1978, 1980b; Blaustein and Wade 1978), guinea pigs (Blaustein and Feder 1979b), and bonnet monkeys (MacLusky et al. 1980). This estrogen-priming effect is also observed in the ring dove (Balthazart et al. 1980).

The induction of progestin receptors by estrogen is dose-dependent in the HPOA and hypophysis of 27-day-old female rats (Kato 1978) and 7-day-old female rats (Kato and Onouchi 1983a). As shown in Fig. 12, with increasing doses of estradiol benzoate the receptor concentrations in the HPOA and hypophysis increase in a linear fashion, and become maximal at doses of 1 and 10 µg, respectively (Kato 1978). It is interesting that these saturation doses are consistent with the doses of ^{3}H-estradiol sufficient to give maximal uptake by the hypothalamus and hypophysis (Kato and Villee 1967b).

The hypothalamic and hypophyseal progestin receptors in 25-day-old female rats reach maximum levels 40–48 h after priming with the potent synthetic estrogen, moxestrol (Moguilewsky and Raynaud 1979a). In 7-day-old female rats, after a single subcutaneous injection of diethylstilbestrol, the hypothalamic binding sites peak 16–24 h later, and gradually decrease after 64 h (Kato and Onouchi 1983a).

There is estrogen specificity in the induction of progestin receptors in the hypothalamus and hypophysis: Neither testosterone, nor the potent androgen R1881, nor progesterone was able to induce receptors in these tissues (Moguilewsky and Raynaud 1979a).

The antiestrogen CI-628 inhibits the induction of progestin receptors by estradiol in the HPOA and pituitary (Roy et al. 1979). Although testosterone treatment increases cytosolic progestin receptor levels in the hypothalamus and preoptic area, it does not alter adenohypophyseal receptor levels. Aromatization of testosterone and estrogen receptor translocation are suggested to be important mechanisms preceding the increase in progestin receptor numbers (Krey et al. 1982).

These results provide further evidence that estradiol induces cytosolic progestin receptors in the HPOA and anterior hypophysis.

3.2 Progesterone-Mediated Progestin Receptor Modulation

Target sensitivity to progesterone could result from variations in the concentration of plasma binding proteins (Clark and Peck 1977), changes in progesterone metabolism (Karavolas and Herf 1971; Stupnicka et al. 1977), or changes in progestin receptor concentration in target organs (Balthazart et al. 1980).

Progesterone injection causes a dose-dependent depletion of cytosolic progestin receptors in guinea pig brain (Blaustein and Feder 1979b) and rat brain (Moguilewsky and Raynaud 1979b; Schwartz et al. 1979), and results in a transient accumulation of nuclear progestin receptors in guinea pig brain (Blaustein 1982b; Blaustein and Feder 1980). The dynamics of progestin receptors in the brain cell are thus essentially similar to those of estrogen receptors (see Sect. 2.1.3.). However, 24 h after the injection in guinea pigs, the concentration of nuclear progestin receptors decreases to preinjection levels, and cytosolic progestin receptor levels remain depressed at 50% of preinjection levels (Blaustein 1982b; Blaustein and Feder 1980). This reduced level of cytosolic progestin receptors is attributable to the refractory period subsequent to progesterone injection (Blaustein 1982a). Similar inhibitory effects of progesterone on hypophyseal cytosolic progestin receptors have been demonstrated in vitro by Haug (1979).

Steroidal modulation and/or regulation of progestin receptors in the brain and hypophysis have been shown to be associated with female sexual behavior and gonadotropin release during prepubertal stages and adulthood. This aspect is mentioned in the next chapter (4.). Hormonal modulation of neonatal brain progestin receptors is discussed later (see Chap. 6).

4 The Physiological Role of Brain Progestin Receptors in Behavioral and Neuroendocrine Responses

Estradiol and progesterone act in concert to activate and terminate female sexual behavior, and to induce ovulation (Powers 1972; Morin 1977; McEwen and Parsons 1982). The physiological role of central progestin receptors is implicated in behavioral and neuroendocrine responses.

4.1 The Estrous Cycle

The influence of both estrogen and progesterone secretion is evident in the changes of the cytosolic and nuclear progestin receptor levels in the hypothalamus and hypophysis during the estrous cycle. Seiki et al. (1969) studied the uptake of ^{3}H-progesterone by the hypothalamus and pituitary of rats at different stages of the estrous cycle. The mean values of radioactivity in all parts of the hypothalamus were highest in proestrus and lowest in estrus; this cyclic variation was statistically significant. As a result of rising estrogen levels in the blood, cytosolic progestin receptor levels in the mediobasal HPOA and hypophysis rise between diestrus and proestrus (McGinnis et al. 1981 a). The increased progesterone level at proestrus causes a rise in the number of nuclear progestin receptors (Rainbow et al. 1982 a). The changes in cytosolic and nuclear progestin receptors in the hypothalamus and hypophysis may be associated with the induction of ovulation and the onset of behavioral estrus.

4.2 Female Sexual Behavior

In the cycling female rat estrogen levels rise for about 48 h, conditioning the neural structure for lordosis, and a surge of progesterone is observed a few hours before the onset of sexual receptivity (induction) (Morin 1977). Termination of sexual receptivity by progesterone follows (a sequential inhibition) (Morin 1977). Simultaneous administration of progesterone and estrogen causes a concurrent inhibition of lordosis (Feder and Marrone 1977; Morin 1977). In the ovariectomized female rat, both facilitative and inhibitory actions of progesterone can be mimicked by estrogen and progesterone (Boling and Blandau 1939; Dempsy et al. 1936; Lisk 1960 a, b; Meyerson 1972; Kubli-Garfias and Whalen 1977; Blaustein and Wade 1977 a, b).

The biochemical mechanisms by which estrogen and progesterone can induce and inhibit female sexual behavior have been analysed at the receptor level. Moguilewsky and Raynaud (1979 b) have reported that changes in progestin receptor levels in the rat hypothalamus and anterior hypophysis are related to the facilitative and inhibitory effects of progesterone on female sexual behavior. The dual action of progesterone seems to be related to an initial estrogen-dependent increase in progestin receptors in the brain, followed by an action of progesterone which results in the reduction of its own receptors. Parsons et al. (1980) have shown the

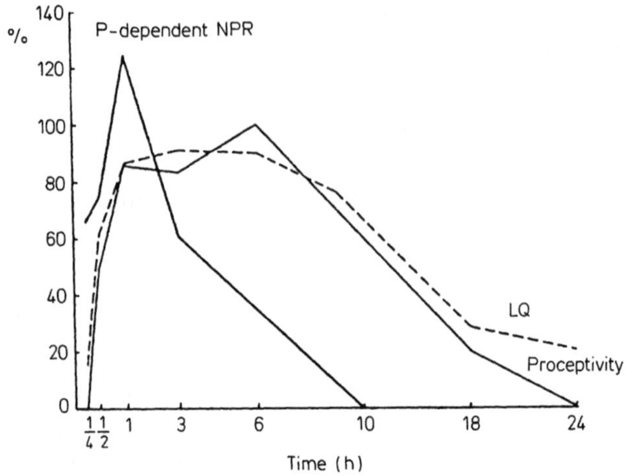

Fig. 13. Comparison between proceptivity and lordosis quotient (LQ) scores, and progesterone(P)-dependent nuclear ^{3}H-R5020 binding (NPR) expressed as percent elevation above control. [From McGinnis et al. (1981 b)]

temporal correlation of progestin receptor induction in the HPOA with the appearance of mating behavior in the female rat. A similar relationship between hypothalamic progestin receptors and the induction of lordosis has been shown to exist in the guinea pig (Blaustein and Feder 1979 a). Moreover, it is interesting that in males the increase in the hypothalamic progestin receptors after estrogen priming is lower than in females (Moguilewsky and Raynaud 1979 b). This provides a clue as to why the doses of estrogen in estrogen-progesterone treatment must be large in order to induce lordosis in the male rat (Pfaff and Zigmond 1971; Södersten 1976; Whalen et al. 1971).

According to McGinnis et al. (1981 b), after a behaviorally effective injection of progesterone, nuclear translocation of progestin receptors in the precedes the earliest appearance of the facilitation of mating behavior (Fig. 13). The behavioral effects of progesterone outlast the nuclear progestin receptor level elevation by several hours. This provides further evidence for the causal relationsship of progestin receptor dynamics in the brain to the initiation of femeale sexual behavior.

Activation of proceptivity and facilitation of receptivity by progesterone are genomically mediated events (Blaustein and Wade 1978; Rainbow et al. 1980). Anisomycin, a brain RNA synthesis inhibitor, has been shown to block the activation of proceptivity and the facilitation of receptivity when administered immediately prior to progesterone (Rainbow et al. 1980). The identification of specific proteins induced by progesterone will aid in establishing whether gene expression plays an important part in the activation of feminine sexual behavior by progesterone (McGinnis et al. 1981 b).

5 Sex Differences in the Brain Progestin Receptor System

5.1 Adult Rodents

The existence of a sex difference in the facilitation of lordosis by estradiol and progesterone in rodents is well established (Beach 1961; Young 1961). Females are more sensitive than males in this behavioral response.

There are negative data on the neural correlates of this sex difference in brain estradiol uptake and concentration of estradiol receptors (Korach and Muldoon 1974; Maurer 1974; Maurer and Wooley 1974; Anderson et al. 1973). On the other hand, there are also positive data on this subject (Whalen and Massicci 1975; Marrone and Feder 1977; DeBold 1978; Whalen and Olsen 1978). There are reports of no sex difference in ^{3}H-progesterone binding with whole-tissue uptake in the brain (Whalen and Luttge 1971a; Wade and Feder 1972). Moguilewsky and Raynaud (1979b) have reported that hypothalami from gonadectomized male rats have a lower concentration of cytosolic progestin receptors than those from females, and that estradiol causes a smaller increase in the receptor concentration in males than in females. They have suggested that the sex differences seen in the expression of lordosis in rodents may be attributable to sex differences in the brain progestin receptor system. Blaustein et al. (1980), using estradiol benzoate, have also demonstrated a smaller increase in the concentration of cytosolic progestin receptors in the hypothalamus, preoptic area, and septum (HPS) in male guinea pigs than in females, although there are no sex differences in receptor concentration in the HPS and cerebral cortex of unprimed male and female animals. Injection of progesterone (0.5 mg) into estrogen-primed males or females results in nuclear progestin receptor levels proportional to cytosolic progestin receptor levels. From these results they conclude that a sex difference in steroid facilitation of lordosis behavior is not sufficiently attributable to a sex difference in brain cytosolic progestin receptors. This may not exclude the possibility that particular neurons which mediate lordosis respond to estrogen priming with induction of cytosolic progestin receptors in females but not in males (Blaustein and Feder 1980).

Recently, Rainbow et al. (1982b), by applying the Palkovits punch-out technique to measure steroid receptors in hypothalamic nuclei, have reported that adult male castrated rats have fewer estrogen and progestin receptors in certain hypothalamic nuclei involved in behavioral and neuroendocrine control than females. The medial preoptic nucleus, which may be involved in estrogen-induced control of gonadotropin release (Pfaff 1980), had less than half the level of estrogen receptors in males as in females. The levels of progestin receptors in the ventromedial nucleus and periventricular region of the preoptic area of estrogen-primed male rats are significantly lower than the receptor levels in estrogen-primed females. The ventromedial nucleus of the hypothalamus may be the principal target site in rat brain for the activation of proceptive and receptive sexual behavior by progesterone (McEwen, cited by Rainbow et al. 1982b). It is interesting that localized differences in estrogen and progestin receptors are not the same in both sexes. These results provide strong evidence for the notion that the sex difference in steroid facilitation of sexual behavior might be ascribable to sex differences in receptor concentrations in some specific hypothalamic nuclei.

5.2 The Early Postnatal Brain

The progestin receptor system in the brain of female rats appears immediately after birth and develops thereafter (MacLusky and McEwen 1980b; Kato and Onouchi 1983a). The postnatal development of the brain receptor system in males is almost the same as in females, in terms of its patterns and concentrations in the HPOA and cortex cytosols (Kato et al. 1984b). However, it is interesting that progestin receptors in purified cell nuclei from the cortex and, probably, HPOA, are fewer in males than in females. The reduced nuclear translocation of the receptors may be at least in part attributable to a sex differenc in the capacity for nuclear receptor translocation. This is supported by the finding that an injection of an excess of progesterone (3 mg) causes less accumulation of progestin receptor complexes in cell nuclei from neonatal male than neonatal female rats (Kato et al. 1984b). It appears that progestin receptor dynamics in the brain during early postnatal life is quite different from that in the adult brain. This sex difference in the receptor system in neonatal rats will be mentioned later in relation to the mechanisms of sexual differentiation of the brain (see Chap. 6).

6 The Ontogeny of the Progestin Receptor System in Brain and Hypophysis and Its Affecting Factors

Responsiveness of the brain to sex steroid hormones alters during its postnatal development (Donovan and van der Werff ten Bosch 1959; Ramirez 1973). Since sex steroid hormones which are localized in specific brain areas are believed to play an essential role in the mechanisms of action of the hormones on the brain, receptor analysis would be a useful tool for elucidation of the molecular mechanism of alterations in brain responsiveness to steroid hormones during development (McEwen 1978a; Kato 1979; MacLusky and Naftolin 1981).

The fetal and postnatal development of the estrogen receptor system in the hypothalamus, cerebral cortex, and hypophysis has been well documented in rats and mice (Kato et al. 1971, 1974; Plapinger et al. 1973; Attardi and Ohno 1976; McEwen 1978a; MacLusky et al. 1979). The postnatal development of the estrogen receptor system in rat brain and pituitary is a dynamic and complex process. There are two distinct phases in the development: an initial rapid phase around the time of birth, during which the receptor system appears and then increases in concentration; and the second, more prolonged phase of differentiation, during which estrogen receptors become differentially distributed within the brain; by days 15–21 the distribution resembles that seen in the adult brain (MacLusky et al. 1979).

Estrogen-concentrating neurons have been found by autoradiography to be concentrated in the hypothalamus, preoptic area and amygdala of the 3-day-old rat (Sheridan et al. 1974a, b). The appearance of the estrogen receptor system in female rat hypothalamus during the "critical" period may be related to hypothalamic sexual differentiation (McEwen 1978a, Lieberburg et al. 1979, MacLusky and Naftolin 1981), although other factors such as androgen receptors may also participate (Kato 1979). Substantial numbers of estrogen receptors are present in the cerebral cortex of the neonatal rat. They increase parallel with hypothalamic estrogen receptors in the first 2 postnatal weeks, and then disappear during the 3rd week of life (Barley et al. 1974). The cortical estrogen receptors in neonates may also be associated with the sexual differentiation process (Barley et al. 1974).

Only a few reports have been available on progestin binding or receptors in the developing brain. Presl et al. (1975) studied ^{3}H-progesterone uptake by the brain and pituitary in newborn, 5-, 10-, 15-, 20-, 25-, and 50-day-old female rats, 1 h after a single injection of ^{3}H-progesterone. High uptake by all tissues in newborn and 5-day-old animals was followed by a sharp decline during the first 2–3 postnatal weeks. Preferential uptake of ^{3}H-progesterone was observed only in the posterior part of the adult rat hypothalamus. On the basis of the lack of a specific uptake mechanism for progesterone in immature females, these investigators assumed that specific progesterone receptors were lacking in the brain regions studied until sexual maturity. Interestingly, in contrast to the hypothalamus, the ratio of radioactivity concentration in the pituitary to that in the cerebellar cortex was already significantly greater than unity at birth, suggesting an earlier appearance of progesterone receptors in the pituitary. More recently, progestin receptors in the neonatal and postnatal brain have been identified (MacLusky and McEwen 1980b; Kato and Onouchi 1981, 1983a) and the postnatal development of the receptor system has been reported (MacLusky and McEwen 1980b; Kato and Onouchi 1983a).

6.1 Identification of Cytosolic Progestin Receptors in the Neonatal Brain

MacLusky and McEwen (1980b) have identified progestin-specific, high-affinity low-capacity ^{3}H-R5020–binding proteins – progestin receptors – in the cytosols rom the hypothalamus and cerebral cortex of the 5-day-old female rat, using gel filtration on LH-20. Kato and Onouchi (1981) have demonstrated that cerebral cortical cytosol from 7-day-old female rats contains a considerable amount of a high-affinity-low-capacity, specific-progestin-binding macromolecular protein with a sedimentation coefficient of $7S$, the characteristics of which are similar to that found in the adult rat brain (Kato and Onouchi 1977; Moguilewsky and Raynaud 1977, 1979a; Kato 1978; Kato et al. 1978; Greenstein 1978; MacLusky and McEwen 1978; Lee et al. 1979). Based on gradient centrifugation and saturation analysis on dextran-coated charcoal, Kato and Onouchi (1983a) have further revealed that ^{3}H-R5020 binding components in the hypothalamic and cortical cytosols of neonatal and developing female rats sediment in the $7S$ region, with a K_d on the order of $10^{-10} M$ and low capacity. These results demonstrate the existence of progestin receptors in the cytosols from neonatal female rat brain.

6.2 Ontogeny of the Progestin Receptor System in Rat Brain and Hypophysis

6.2.1 Cytosolic Progestin Receptors

6.2.1.1 Female Rats

The ontogeny of the progestin receptors in various parts of the brain and hypophysis of intact female rats has been reported at successive developmental stages, from fetal to postnatal (MacLusky and McEwen 1980b; Kato and Onouchi 1983a). Cytosolic progestin receptors in the brain and hypophysis were determined by single point assay on Sephadex LH-20 (MacLusky and McEwen 1980b), and by saturation analysis on dextran-coated charcoal (Kato and Onouchi 1983a).

Progestin receptors seem to be lacking in brain and hypophyseal cytosols from fetuses at 19 and 21 gestational days. At birth (day 0), the cytosolic receptors are barely detectable in the hypothalamus-preoptic area (HPOA) and midbrain. A low level of receptor in the cortex was detected by Kato and Onouchi (1983a), but not by MacLusky and McEwen (1980a). At day 1, a low but definite level of progestin receptor is found in the HPOA, cortex, amygdala, and cerebellum. Thus, putative progestin receptors in these brain areas appear immediately after birth. The receptor level increases slightly in the HPOA, cerebral cortex,

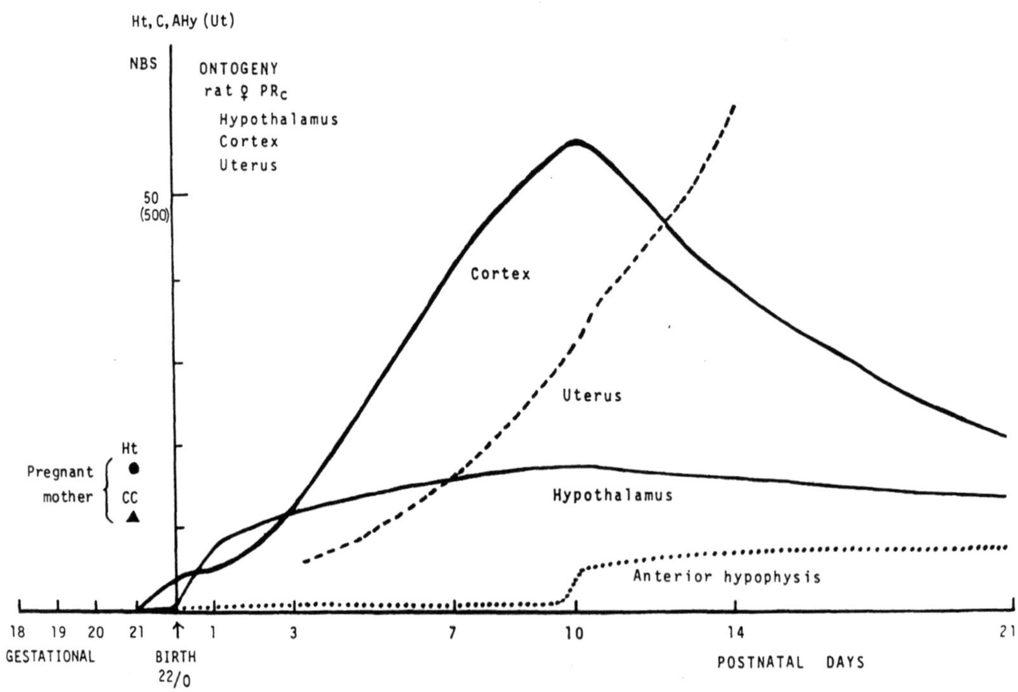

Fig. 14. The fetal and postnatal development of the cytosolic progestin receptor (PRc) system in female rat brain, hypophysis and uterus. Ht, Hypothalamus (hypothalamus-preoptic area); CC, cerebral cortex; C, cerebral cortex; AHy, anterior hypophysis; Ut, uterus; NBS, number of binding sites [Data derived from Kato and Onouchi (1983a)]

amygdala, cerebellum, and midbrain at day 3, followed by a further increase at day 7. Thereafter, the development of progestin binding becomes different in the HPOA and cerebral cortex. As shown in Fig. 14, the concentration of cytosolic progestin receptors in the HPOA remains at almost the same level throughout the postnatal period examined. In contrast, the cytosolic progestin receptors in the cerebral cortex increase rapidly at 7–10 days of age, reach a maximal level at day 10, and decrease to a low level by days 21–28.

The development of progestin receptors in female rats is different in the hypophysis than in the brain (MacLusky and McEwen 1980a; Kato and Onouchi 1983a). Hypophyseal ^{3}H-R5020 binding is undetectable during the 1st postnatal week and remains low through the 3rd week, when a rapid increase follows during the late prepubertal stage at 28 days of age. Receptor levels remain low until 25 days of age (MacLusky and McEwen 1980a). An increase in blood estrogen may be responsible for the elevation of progestin receptor levels, because the hypophysis is very responsive to estradiol induction of progestin receptors (Kato and Onouchi 1977; Moguilewsky and Raynaud 1977, 1979a; Kato et al. 1978; MacLusky and McEwen 1978).

6.2.1.2 Male Rats

Accordingly to our recent data (Kato et al. 1984b), the fetal and postnatal development of progestin receptor systems in the cytosols of the HPOA and cerebral cortex of male rats was essentially similar to that in female rats (Fig. 15).

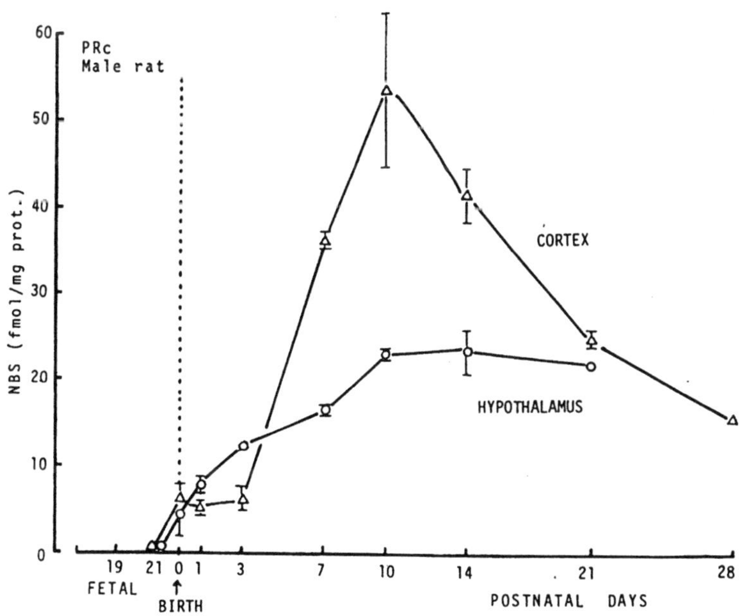

Fig. 15. The fetal and postnatal development of the cytosolic progestin receptor (PRc) system in male rat hypothalamus and hypophysis. NBS, number of binding sites. [Data derived from Kato et al. (1984b)]

It is interesting that some male-female differences were observed in the HPOA at days 10–14; the receptor concentration in males was greater than that in females.

6.2.2 The Onset of Nuclear Receptor Translocation

As previously stated, the cytosolic progestin receptors in rat brain appear immediately after birth then progressively develop, reaching a maximum level at around day 10. Thereafter, the hypothalamic receptors remain at almost a constant level, and the cortical receptors decline progessively to the adult level at days 21–28. In this connection, the levels of plasma progesterone during development are interesting. According to Döhler and Wuttke (1974), the level of plasma progesterone in female rats is low during the first 10 days, slowly increases between days 10–20, and progessively increases during the 3rd and 4th weeks. Meijs-Roelofs et al. (1975, 1981) have reported that plasma progesterone is detected at day 13 and reaches a maximum at days 18–22, and that progesterone increase at day 20 is associated with the control of FSH secretion. In this context, it is very interesting to learn about nuclear translocation of brain progestin receptors during development.

6.2.2.1 The Postnatal Development of Nuclear Progestin Receptors in Female Rat Brain

Kato and Onouchi (1983 a) have reported nuclear progestin receptors in purified cell nuclei from the HPOA and cerebral cortex of neonatal and developing female rats. The number of binding sites was determined in purified nuclei by low-temperature-exchange assay (Blaustein and Feder 1980). The developmental patterns of nuclear progestin receptors in the cerebral cortex and HPOA of intact female rats are shown in Fig. 16. Cortical nuclear progestin binding is very low until day 3, but dramatically increases at days 7–10, remaining high thereafter. This may indicate a sudden onset of increased activation and/or nuclear translocation of endogenous progestin-receptor complexes at days 7–10, despite the low level of serum progesterone at this time (Döhler and Wuttke 1974; Meijs-Roelofs et al. 1975, 1981). A similar though less pronounced pattern is seen in the HPOA nuclei. Moreover, when progesterone was given to 10-day-old female rats, the progesterone-receptor complex was translocated into cell nuclei (Kato et al. 1984 b). Increased nuclear translocation of endogenous progestin-receptor complexes may occur in the cortex and, possibly, HPOA at around days 7–10. This sudden increase of nuclear translocation is probably not due to a change in progesterone levels, serum progesterone levels are low and constant during early development. The cause and physiological significance of the increase is entirely unknown.

6.2.2.2 The Developmental Pattern of Nuclear Progestin Receptors in the Male Rat and Sex Differences in Nuclear Translocation

Kato et al. (1984 b) have shown that the developmental pattern of nuclear receptor translocation in male rat brain essentially resembles that in females. However,

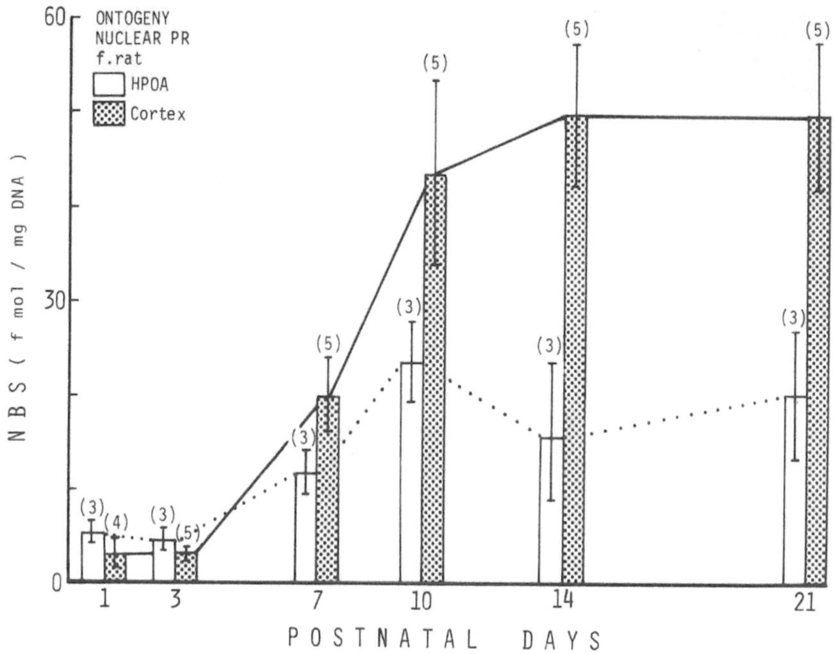

Fig. 16. Nuclear progestin binding in the female rat hypothalamus and cerebral cortex during postnatal development. Extract from the hypothalamic or cerebral cortical purified nuclei was incubated with 4 nM ^{3}H-R5020 in the presence or absence of excess unlabeled R5020. Number of binding sites (NBS) was expressed as fmol per mg DNA. Each value represents the mean ± SE (vertical bar). The number in parenthesis indicate the number of determinations in each group. [From Kato and Onouchi (1983 a)]

less binding in the cortex and, possibly, HPOA was found in males at days 10–21. After administration of progesterone, the developing male rats accumulated a lower concentration of progestin receptors in the cortex and, possibly, HPOA than similarly treated females (Fig. 17).

The conclusion was that there is a marked sex difference in nuclear progestin receptor systems in the postnatal rat brain, particularly in the cortex. The postnatal male rat brains seems to have lower capacities for nuclear receptor translocation than the female one. The progestin receptor system in early life might be involved in the mechanism of sexual differentiation of the rat brain.

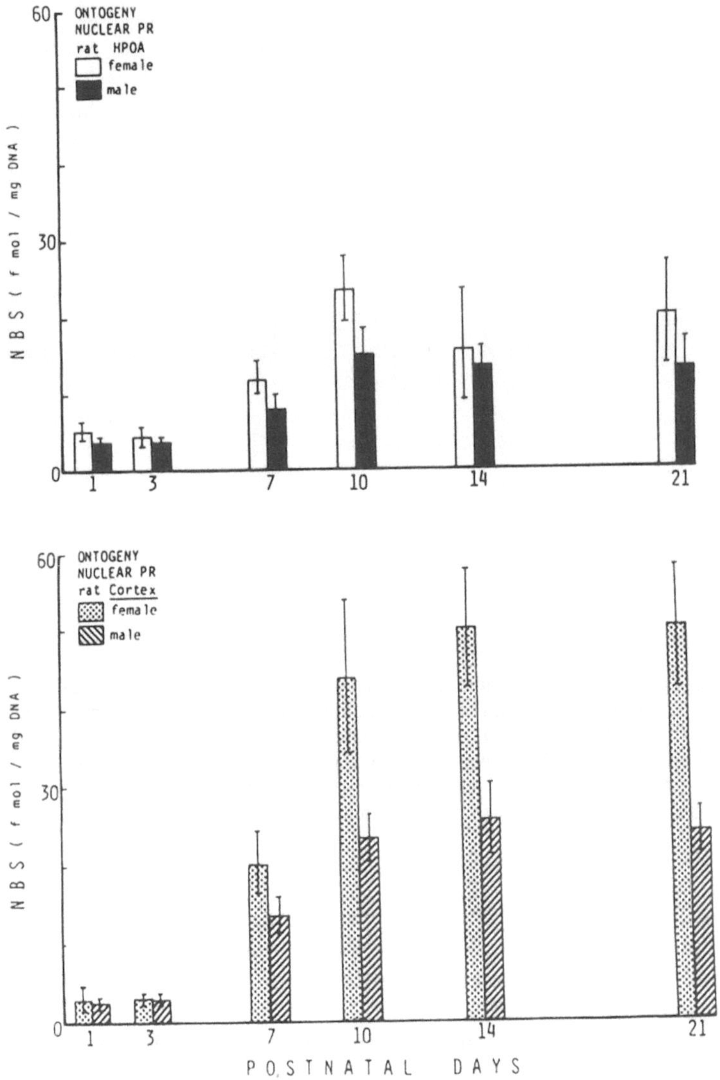

Fig. 17. A sex difference in nuclear progestin receptors (PR) in the rat hypothalamus and cerebral cortex. HPOA, Hypothalamus-preoptic area; cortex, cerebral cortex; NBS, number of binding sites expressed as fmol per mg DNA [Data derived from Kato and Onouchi T (1983 a); Kato et al. (1984 b)]

6.3 Hormonal Factors Affecting Progestin Receptors in the Developing Brain

6.3.1 Effects of Estrogen Priming

The HPOA and hypophysis in the developing rat are sensitive to estrogen induction of progestin receptors, but the cerebral cortex is not (MacLusky and McEwen 1980 a). It has been postulated that the development of mechanisms for estrogen-dependent behavioral and neuroendocrine responses to progesterone may be associated with maturation of the estrogen-inducible progestin receptor system within the HPOA and hypophysis (MacLusky and McEwen 1980a). These findings have been confirmed and extended by a recent report (Kato and Onouchi 1983a).

6.3.1.1 Different Effects of Estrogen Priming

Gradient centrifugation studies have revealed different age-dependent effects of diethylstilbestrol (DES), on the induction of 7*S* progestin receptors in the hypothalamus but not in the cortex, of the estrogen-treated female rat. Estradiol had no definite effect on the progestin binding peak in the 7- and 14-day-old rat hypothalamus (Kato and Onouchi 1983a). The failure of estradiol to induce hypothalamic progestin receptors might be ascribed to the presence of fetoneonatal estradiol-binding protein (fEBP) in the hypothalamus (Plapinger et al. 1973; Kato et al. 1974; McEwen 1978a).

When dose dependency of estrogen-inducible cytosolic progestin receptors in the neonatal brain was examined, the hypothalamic progestin binding capacity increased with a 0.1 μg dose of DES, and reached a maximum with 1 μg or more. Less inductive effect was found with estradiol; at 10 μg the progestin binding capacity was increased, and at 50 μg, saturated. In contrast, the cerebral cortical binding sites seemed affected by neither DES nor estradiol at doses up to 1 μg. There seem to be differential inductive effects of estrogen in the cortex and hypothalamus. It is noteworthy that large doses of estrogens caused a degradative effect on the brain receptors.

6.3.1.2 Time Course of Progestin Receptor Induction

Following a single injection of DES, hypothalamic progestin receptors were induced 16–24 h later. Cortical progestin binding sites were not affected by DES.

The above results suggest that the neonatal hypothalamus and hypophysis but not the cortex, are sensitive to DES induction of progestin receptors.

6.3.2 Progesterone Modulation

Comparison of the early postnatal development of the estrogen and progestin receptor systems in the brain is interesting. Estrogen and progestin receptors increase rapidly to reach a peak level at day 7, but the estrogen receptor level declines precipitously at day 14, while the progestin receptor levels remain high at this age. High cortical levels of progestin receptor, and enhanced nuclear translocation at around day 10, imply some function of cortical progestin receptors in the regulatory mechanisms of the cortical estrogen receptor, although the physiological significance of the cortical progestin receptor is still unkown.

In this context, it is interesting learn how steroid receptors in developing brain are modulated and/or regulated by progesterone. Progesterone can depress the level of cytosolic estrogen receptors in the adult rat uterus (Hsueh et al. 1976; Aihara et al. 1980). The effects of progesterone on estrogen receptors on the adult rat brain are, however, still controversial and conflicting (Reuter and Lisk 1976; Pavlik and Coulson 1976; DeBold et al. 1976; Lisk and Reuter 1977; Marrone and

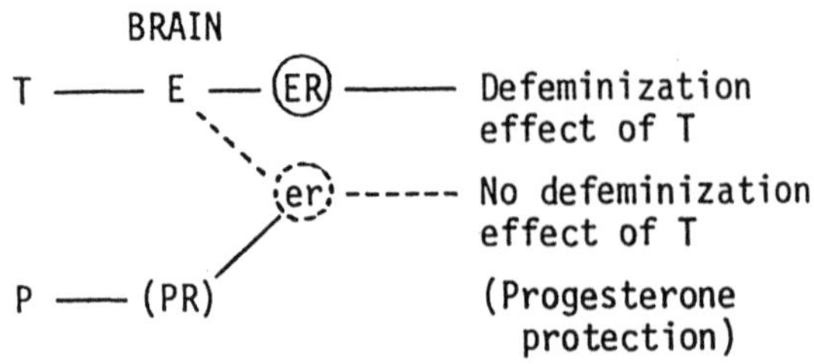

Fig. 18. A possible mechanism of the protective action of progesterone against the neural defeminization effect of exogenous androgen in the neonatal female rat. T, testosterone; E, estrogen; P, progesterone; PR, progesterone receptor; ER, estrogen receptor; er, lowered level of estrogen receptor

Feder 1977; Haug 1979; Schwartz et al. 1979; Attardi 1981; Kato and Onouchi 1983b). Little information is available about the effects of progesterone on estrogen receptors in the neonatal rat brain. Kato and Onouchi (1984) have examined the effects of progesterone on estradiol receptor levels in the HPOA, cerebral cortex, and uterus of 7-day-old female rats. They used ^{3}H-moxestrol which does not bind fEBP. Progesterone (3 mg) lowered the concetrations of total and cytosolic estradiol receptors in the HPOA and uterus, but not in the cortex. The HPOA estrogen receptor levels may be affected by progesterone through the progesterone receptors which apppear in the neonatal rat brain. The postnatal development of estrogen receptors in the cortex seems to be independent of progesterone effects.

The inhibitory effect of progesterone on neonatal rat estrogen receptors is interesting with respect to its protective action, in female rats, against the neural defeminization effects of exogenous androgen (Kincl and Maqueo 1965; Cagnoni et al. 1965; Arai and Gorski 1968; Gonzalez-Mariscal et al. 1982). The molecular mechanisms underlying this protective action are unkown. Gonzalez-Mariscal et al. (1982) have proposed that anesthetic properties of progestins counteract the "organizing" action of androgen upon the newborn rat brain. They further assume that progestins counteract defeminization by modifying the insertion or secretion of the androgen-induced macromolecules which mediate anatomical changes in the brain. The level of neonatal rat hypothalamic estrogen receptors is reduced by 3 mg progesterone, but not by 1 mg (Kato and Onouchi, unpublished data). The progersterone doses effective on neonatal brain estrogen receptors are roughly consistent with those which protect against the defeminization effect of exogenous androgens in the neonatal female rat (Arai and Gorski 1968; Gonzalez-Mariscal et al. 1982). The inhibitory effect of progesterone on hypothalamic estrogen receptors may be attributable to its anesthetic property rather than its progestational nature. However, this effect is evident only in the hypothalamus, not in the cortex; indicating its selective action on the neonatal brain estrogen receptors. It is thus possible that progesterone counteracts virilization by its inhibitory effect upon estrogen receptors in the hypothalamus of neonatal rats (Fig. 18).

6.3.3 Thyroid Hormone as a Possible Affecting Factor

Kikuyama (1969) has reported that hypothyroidism at birth prolongs the critical period if female rat susceptibility to the defeminizing effects of testosterone propionate (TP). Hyperthyroidism shortly after birth, on the other hand, shortens the critical period of susceptibility to TP (Phelps and Sawyer 1976). Hypothyroidism and hyperthyroidism in the early postnatal period affect myelination and development of the neuropil, and synaptogenesis (McEwen 1978 b). Thyroid hormone is involved in neural maturation and sexual differentiation. There are no data available on its relationship to steroid sex hormone receptors in the brain.

Kato et al. (1984 a) have reported that the hypothyroid state induced by 6-*n*-propyl-2-thiouracil (PTU) injections lowers 'selectively' the cerebral cortical progestin receptor level in 10-day-old rats. These results suggest that thyroidal factor(s) are involved in ontogenesis of the progestin receptor system of the rat cortex.

6.3.4 Summary

The fetal and postnatal development of the progestin receptor system in the brain and hypophysis is a dynamic and complex process. The cytosolic receptors in the cerebral cortex rapidly increase, with a peak at days 7–10 followed by a decline; the hypothalamic receptors remain at a stationary level from day 7 onwards. The developmental patterns are essentially similar in males and females. With regard to nuclear receptors, there is a sudden onset of increased nuclear translocation and/or activation of the progestin receptor system in the cortex and, possibly, hypothalamus, around days 7–10. A marked sex difference exists in the postnatal development of nuclear receptor translocation. The male brain seems to have lower capacities for nuclear binding than the female one. The postnatal brain progestin receptor system is implicated in the mechanisms of sexual differentiation of the brain.

In the developing brain, ovarian steroids modulate their receptors in some specific brain areas (HPOA and hypophysis) but not in the cortex. The postnatal development of the cortical progestin receptor system seems to be independent of steroidal regulation. The regulatory mechanism for the hypothalamic receptors in development is still unknown, although steroidal modulation is possible. Thyroid hormone is one possible candidate as a factor affecting the development of the brain progestin receptor system, because neonatal hypothyroidism can selectively depress the receptor level.

7 Summary and Conclusion

Utilization of ^{3}H-R5020, a synthetic progestin, as a specific ligand for progestin receptor has made it possible to unequivocally demonstrate the presence of high-affinity–low-capacity specific progestin-binding macromolecules in the cytosols of the brain and hypophysis of various kinds of mammals and some birds. Nuclear progestin receptors have been also identified in the brain and hypophysis. Physico-chemical properties of the cytosolic and nuclear progestin receptors in the brain and hypophysis are similar to those of other steroid receptors in the tissues. It is noteworthy, however, that an in vitro exchange reaction of R5020 or progesterone in the brain and hypophyseal cytosols and nuclei is completed at low temperatures (0–10 °C), in contrast to the exchange reaction of estradiol which occurs at higher temperatures (25–37 °C).

There are two kinds of progestin receptors in the brain: estrogen-inducible and -uninducible receptors. Estrogen can induce progestin receptors dose dependently in some specific areas such as the HPOA and anterior hypophysis. However, cortical progestin receptors are not modulated by estrogen. The two kinds of receptors are similar in character. Cytosolic receptor-progestin complexes can be translocated into nuclei. Unlike estradiol receptors, cytosolic progestin receptors are not replenished. It is interesting that both estrogen-inducible and -uninducible progestin receptors in the hypothalamus and cerebral cortex are translocated in the same ratio into cell nuclei of adult and neonatal rats after progesterone injection. The physiological function of uninduced progestin receptors has not yet been explained.

Gross biochemical, autoradiographic, and mapping studies in the rodent and primate brain have revealed that estrogen-inducible progestin receptors are distributed similarly estrogen receptors. Brain progestin receptors in the rat are more widely distributed than in the monkey.

Progestin receptor-containing nuclei and brain regions may be interconnected by neural pathways. It is particularly of interest that the amygdala contains few if any progestin receptors, even in the estrogen-primed rat and monkey, despite relatively high levels of estrogen receptors in the area. This finding indicates that amygdala progestin receptors are not inducible by estrogen. The brain and hypophysis of the monkey and, possibly, rodent can be categorized, in terms of estrogen receptors (ER) and progestin receptors (PR), into ER+PR+ (HPOA and anterior hypophysis), ER+PR− (amygdala) and ER−PR± (cerebral cortex and cerebellum) groups. Progesterone may act differentially on the brain. This suggests a possible central action of progesterone through the progestin receptors in the HPOA. A deficient or low level of progestin receptors in the amygdala suggests that progesterone does not act directly on the amygdala, especially in the monkey. Further consideration might be given, however, to a direct but non-receptor-mediated action or an indirect effect of progesterone on the amygdala.

The biochemical mechanisms by which progesterone can activate and terminate female sexual behavior and induce ovulation have been analyzed on a receptor level. There is a causal relationship between progestin receptor dynamics in the brain and female sexual behavior or neuroendocrine function. The physiological role of brain progestin receptors is implicated in behavioral and neuroendocrine responses. It is very interesting that activation of proceptivity and facilitation of receptivity by progesterone are genomically mediated events. The identification of progesterone-inducible specific proteins is necessary for an understanding of the biochemical processes.

There are sex differences in the concentrations of progestin receptors in some specific hypothalamic nuclei of the rodent brain. The receptor level in some specific hypothalamic nuclei is lower in males than in females. This is in agreement with the observation that females are more sensitive than males in their behavioral response. The sex difference might be ascribable to the sex difference in receptor concentrations in some specific hypothalamic nuclei.

The fetal and postnatal development of the progestin receptor system in the rat brain and hypophysis is a dynamic and complex process. The brain cytosolic receptors appear immediately after birth and rapidly increase in early neonatal stages with a peak at days 7–10; the level in the cerebral cortex then declines. The hypothalamic receptors maintain a stationary level from day 7 onwards. The hypophyseal receptors develop at later prepubertal stages. The developmental changes in the receptor system may be related to development of the neuroendocrine functions and sexual behavior. Despite the low level of blood progesterone, there is a sudden onset of increased nuclear translocation and/or activation of the progestin receptor system in the cortex and, possibly, hypothalamus at around days 7–10. The physiological role of this phenomenon is entirely unknown. A marked sex difference exists in the postnatal development of nuclear receptor translocation. The male brain seems to have lower capacities for nuclear binding than the female brain. The postnatal brain progestin receptor system is implicated in the mechanisms of sexual differentiation of the brain. While steroidal modulation and/or regulation of progestin receptors is implicated in the developing hypothalamus, the postnatal development of cortical receptors seems to be independent of steroidal modulation. Thyroid hormone may be a candidate as one of the factors affecting the development of the progestin receptor system in the brain.

Acknowledgements. The author expresses his sincere appreciation to Prof. S. Okinaga, Prof. K. Arai, Prof. K. Oshima, Dr. T. Onouchi, and Ms. M. Takamatsu for their generous assistance and to Mrs. Y. Onda for preparation of the manuscript. He also thanks Prof. D. W. Pfaff of Rockefeller University for reviewing the manuscript. This work was supported in part by grants from the Ministry of Education, Science and Culture, The Japanese Government (No. 5612006, No. 57113007, No. 5780314, 58105007, No. 59370001 and No. 59370016).

References

Aihara M, Kimura T, Kato J (1980) Dynamics of the estrogen receptor in the uterus of mice treated neonatally with estrogen. Endocrinology 107:224–230

Anderson JN, Clark JH, Peck EJ Jr (1972) Oestrogen and nuclear binding sites. Determination of specific sites by ^{3}H-oestradiol exchange. Biochem J 126:561–567

Anderson JN, Peck EJ Jr, Clark JH (1973) Nuclear receptor estrogen complex. Accumulation, retention, and localization in the hypothalamus and pituitary. Endocrinology 93:711–717

Arai Y, Gorski R (1968) Protection against the neural organizing effect of exogenous androgen in the neonatal female rat. Endocrinology 82:10005–1009

Atger M, Baulieu EE, Milgrom E (1974) An investigation of progesterone receptors in guinea pig vagina, uterine cervic, mammary glands, pituitary and hypothalamus. Endocrinology 94:161–167

Attardi B (1981) Facilitation and inhibition of the estrogen induced luteinizing hormone surge in the rat by progesterone: effects on cytoplasmic and nuclear estrogen receptors in the hypothalamic preoptic area, pituitary, and uterus. Endocrinology 108:1487–1496

Attardi B, Ohno S (1976) Androgen and estrogen receptors in the developing mouse brain. Endocrinology 99:1279–1290

Balthazart J, Blaustein JD, Cheng MF, Feder HH (1980) Hormones modulate the concentration of cytoplasmic progestin receptors in the brain of male ring doves (*Streptopelia risoria*). J Endocrinol 86:251–261

Barley J, Ginsburg M, Greenstein BD, MacLusky NJ, Thomas PJ (1974) A receptor mediating sexual differentiation? Nature 252:259–260

Barraclough CA (1973) Sex steroid regulation of reproductive neuroendocrine processes. In: Greep RO, Astwood EB (eds) Handbook of physiology, Sect. 7. Endocrinology vol. 2 Female Reproductive System, Part I. American Physiological Society, Washington DC, pp 29–56

Beach FA (1961) Hormones and behavior. Cooper Square, New York

Blaustein JD (1982 a) Progesterone in high doses may overcome progesterone's desensitization effect on lordosis by translocation of hypothalamic progestin receptors. Horm Behav 16:175–190

Blaustein JD (1982 b) Alteration of sensitivity to progesterone facilitation of lordosis in guinea pigs by modulation of hypothalamic progestin receptors. Brain Res 243:287–300

Blaustein JD, Feder HH (1979 a) Cytoplasmic progestin receptors in guinea pig brain. Characteristics and relationship to the induction of sexual behavior. Brain Res 169:481–497

Blaustein JD, Feder HH (1979 b) Cytoplasmic progestin receptors in the female guinea pig brain and their relationship to refractoriness in expression of female sexual behavior. Brain Res 177:489–498

Blaustein JD, Feder HH (1980) Nuclear progestin receptors in guinea pig brain measured by an in vitro exchange assay after hormonal treatments that affect lordosis. Endocrinology 106:1061–1069

Blaustein JD, Wade GN (1977 a) Concurrent inhibition of sexual behavior, but not brain (^{3}H)estradiol uptake, by progesterone in female rats. J Comp Physiol Psychol 91:742–751

Blaustein JD, Wade GN (1977 b) Sequential inhibitions of sexual behavior by progesterone in female rats: comparison with a synthetic antiestrogen. J Comp Physiol Psychol 91:752–760

Blaustein JD, Wade GN (1978) Progestin binding by brain and pituitary cell nuclei and female rat sexual behavior. Brain Res 140:360–367

Blaustein JD, Ryer HI, Feder HH (1980) A sex difference in the progestin receptor system of guinea pig brain. Neuroendocrinology 31:403–409

Boling JL, Blandau RI (1939) The estrogen-progesterone induction of mating responses in the spayed female rat. Endocrinology 25:359–364

Brown-Grant K, Naftolin F (1972) Facilitation of luteinizing hormone secretion in the female rat by progesterone. J Endocrinol 53:37–46

Burton K (1956) A study of the conditions and mechanism of diphenylamine reaction for the colorimetric estimation of deoxyribonucleic acid. Biochem J 62:315–323

Cagnoni M, Fantini F, Morace G, Ghetti A (1965) Failure of testosterone propionate to induce the "early androgen" syndrome in rats previously injected with progesterone. J Endocrinol 33:527–528

Caligaris L, Astrada JJ, Taleisnik S (1971) Biphasic effect of progesterone on the release of gonadotropin in rats. Endocrinology 89:331–337

Clark JH, Peck EJ (1977) Steroid hormone receptors: basic principles and measurements. In: Receptors and hormone action, vol 1, O'Malley BW, Birnbaumer L (eds). Academic, New York

Clark JH, Eriksson HA, Hardin JW (1976) Uterine receptor-estradiol complexes and their interaction with nuclear binding sites. J Steroid Biochem 7:1039–1043

Clifton DK, Steiner RA, Resko JA, Spies HG (1975) Estrogen-induced gonadotropin release in ovariectomized rhesus monkeys and its advancement by progesterone. Biol Reprod 13:190–194

Danzo BJ, Eller BC (1976) Nuclear binding of [^{3}H]androgens by the epididymis of sexually mature castrated rabbits. J Steroid Biochem 7:733–739

DeBold JF (1978) Modification of nuclear retention of [^{3}H]-estradiol by cells of the hypothalamus as a function of early hormone experience. Brain Res 159:416–420

DeBold JF, Martin JV, Whalen RE (1976) The excitation and inhibition of sexual receptivity in female hamsters by progesterone: time and dose relationships, neural localization and mechanisms of action. Endocrinology 99:1519–1527

Dempsey EW, Hertz R, Young WC (1936) The experimental induction of oestrus (sexual receptivity) in the normal and ovariectomized guinea pig. Am J Physiol 116:201–209

Dierschke DJ, Yamji T, Karsch FJ, Weick G, Weiss G, Knobil E (1973) Blockade by progesterone of estrogen-induced LH and FSH release in the rhesus monkey. Endocrinology 92:1496–1501

Döhler DD, Wuttke W (1974) Serum LH, FSH, prolactin and progesterone from birth to puberty in female and male rats. Endocrinology 94:1003–1008

Donovan BT, van der Werff ten Bosch JJ (1959) The hypothalamus and sexual maturation in the rat. J Physiol 147:78–92

Dörner G (1972) Sexualhormonabhängige Gehirndifferenzierung und Sexualität. Fischer, Jena

Everett JW (1961) The mammalian female reproductive cycle and its controlling mechanisms. In: Sex and internal Secretions, 3rd edn. Young WC (eds). Williams and Wilkins, Baltimore

Feder HH, Marrone BL (1977) Progesterone: its role in the central nervous system as a facilitator and inhibitor of sexual behavior and gonadotropin release. Ann NY Acad Sci 286:331–354

Ferin M, Rosenblatt H, Carmer PW, Antunes JL, Vande Wiele RL (1979) Estrogen-induced gonadotropin surges in female rhesus monkeys after pituitary stalk section. Endocrinology 104:50–52

Freeman MC, Dupke KC, Croteau CM (1976) Extinction of the estrogen-induced daily signal for LH release in the rat: a role for the proestrous surge of progesterone. Endocrinology 99:223–229

Furr BJA, Bonney RD, England RJ, Cunningham FJ (1973) Luteinizing hormone and progesterone in peripheral blood during the ovulatory cycle of the hen (*Gallus domesticus*). J Endocrinol 57:159–169

Garris DR, Billiar RB, Takaoka Y, White RJ, Little B (1981) In situ estradiol and progestin (R5020) localization in the vascularly separated and isolated hypothalamus of Rhesus monkey. Neuroendocrinology 32:202–208

Gerlach JL, McEwen BS, Pfaff DW, Moskovitz S, Ferin M, Carmel PW, Zimmerman EA (1976) Cells in regions of rhesus monkey brain, and pituitary retain radioactive oestradiol, corticosterone and cortisol differentially. Brain Res 103:603–612

Goldman BD, Zarrow MX (1973) The physiology of progestins. In: Handbook of physiology, vol 2, sect 7. Greep RO (ed). American Physiological Society, Washington DC

Gonzalez-Mariscal G, Fernandez-Guasti A, Beyer C (1982) Anesthetic pregnanes counteract androgen-induced defeminization. Neuroendocrinology 34:357–362

Goodman RL (1978) A quantitative analysis of the physiological role of estradiol and progesterone in the control of tonic and surge secretion of LH in the rat. Endocrinology 102:142–150

Greenstein BD (1978) Evidence for specific progesterone respectors in rat brain cytosol. J Endocrinol 79:327–338

Haug E (1979) Progesterone suppression of estrogen-stimulated prolactin secretion and estrogen receptor levels in rat pituitary cells. Endocrinology 104:429–437

Hess DL, Wilkins RH, Moossy J, Chang JL, Plant TM, McCormak JT, Nakai Y, Knobil E (1977) Estrogen-induced gonadotropin surges in decerebrated female rhesus monkeys with medial basal hypothalamic peninsuale. Endocrinology 101:1264–1271

Hsueh AJW, Peck EJ Jr, Clark JH (1976) Control uf uterine estrogen receptor levels by progesterone. Endocrinology 98:438–444

Iramain CA, Danzo BJ, Strott CA, Toft DO (1973) Progesterone binding in the hypothalamus and hypophysis of female guinea pigs and rabbits. Proc IVth Int Congr Int Soc Psychoneuroendocrinol University of California, Berkeley, p 15 (Abstract)

Karavolas JH, Herf SM (1971) Conversion of progesterone by rat medial basal hypothalamic tissue to 5α-pregnane-3, 20-dione. Endocrinology 89:940–942

Karsch FJ, Weick RF, Hotchkiss J, Dierschke DJ, Knobil E (1973) An analysis of the negative feedback control of gonadotropin secretion utilizing chronic implantation of ovarian steroids in ovariectomized rhesus monkeys. Endocrinology 93:478–486

Kato J (1963) Cholinergic and adrenergic mechanism in female rat hypothalamus with special reference to the reproductive functions. Endocrinol Jpn 10:145–146

Kato J (1975) The role of hypothalamic and hypophyseal 5α-dihydrotestosterone, estradiol and progesterone receptors in the mechanism of feedback action. J Stereoid Biochem 6:979–987

Kato J (1977 a) Characterization and function of steroids receptors in the hypothalamus and hypophysis. In: James VHT (ed) Endocrinology, Proc V Intern Congress of Endocrinology, Hamburg, 1976, vol 1. Excerpta Medica, Amsterdam, pp 12–17 (Abstract)

Kato J (1977 b) Steroid hormone receptors in brain, hypothalamus and hypophysis. In: Pasqualini JP (ed) Receptors and mechanisms of action of steroid hormones, part II. Dekker, New York, pp 603–671

Kato J (1978) Brain receptors for sex steroid hormones and their characterization and functional implications. In: Scott DE, Kozlowski GP, Collins F (eds) Brain-Endocrine Interaction III. Neural hormones and reproduction. Karger, Basel, pp 286–301

Kato J (1979) Steroid hormone receptors during brain development. In: Kaye AM, Kaye M (eds) Advances in the biosciences, vol 25. The development of responsiveness to steroid hormones. Pergamon, Oxford, pp 389–413

Kato J, Onouchi T (1977) Specific progesterone receptors in the hypothalamus and anterior hyphophysis of the rat. Endocrinology 101:920–928

Kato J, Onouchi T (1979) Nuclear progesterone receptors and characterization of cytosol receptors in the rat hypothalamus and anterior hypophysis. J Steroid Biochem 11:845–854

Kato J, Onouchi T (1981) Progesterone receptors in the cerebral cortex of neonatal female rats. Dev Neurosci 4:427–432

Kato J, Onouchi T (1983 a) Progestin receptors in female rat brain and hypophysis in the development from fetal to postnatal stages. Endocrinology 113:29–36

Kato J, Onouchi T (1983 b) Effects of ovarian steroid hormones on the brain and hypophysis: Receptor modulation and chromatin binding. In: McKerns KW (ed) Regulation of gene expression by hormones. Plenum, New York, pp 221–234

Kato J, Onouchi T (1984) Ovarian steroid hormone receptors in developing rat brain and progesterone modulation of estrogen receptors in the hypothalamus. In: Saxena BB, Catt KJ, Birnbaumer L, Martini L (eds) Hormone receptors in growth and reproduction, Serono Symposia Publications, vol. 9, 1982. Raven, New York, pp 279–289

Kato J, Villee CA (1967 a) Preferential uptake of estradiol by the anterior hypothalamus of the rat. Endocrinology 80:1133–1138

Kato J, Villee CA (1967 b) Factors affecting uptake of estradiol-6,7-^{3}H by the hypophysis and hypothalamus. Endocrinology 80:1133–1138

Kato J, Atsumi Y, Inaba M (1971) Development of estradiol receptors in rat hypothalamus. J Biochem 70:1051–1053

Kato J, Atsumi Y, Inaba M (1974) Estradiol receptors in female rat hypothalamus in the developmental stages and during pubescence. Endocrinology 94:309–317

Kato J, Onouchi T, Okinaga S (1978) Hypothalamic and hypophysial progesterone receptors: estrogen priming effect, localization and isolation of nuclear receptors. J Steroid Biochem 9:419–427

Kato J, Onouchi T, Oshima K (1980) The presence of progesterone receptors in the sexual skin of the monkey. Steroids 35:743–749

Kato J, Onouchi T, Oshima K (1981) Brain receptors for progesterone and estrogen in the Japanese monkey (*Macaca fuscata*) and possible differential central actions of progesterone. In: Cortes-Prieto J, Campos da Paz A, Neves-e-Castro M (eds) Research on fertility and sterility. MTP Press, Lancaster, England, pp 213–220

Kato J, Onouchi T, Takamatsu M (1984 a) Decreased progestin receptors in the cerebral cortex of hypothyroid postnatal rats. J steroid Biochem 20:817–820

Kato J, Onouchi T, Okinaga S, Takamatsu M (1984 b) The ontogeny of cytosol and nuclear progestin receptors in male rat brain and its male-female differences. J Steroid Biochem 20:147–152

Katzenellenbogen JA, Johnson HJ, Carlson KE (1973) Studies on the uterine, cytoplasmic estrogen binding protein; thermal stability and ligand dissociation in rats. An assay of empty and filled sites by exchange. Biochemistry 12:4092–4099

Kawashima M, Kamiyoshi M, Tanaka K (1978) A cytoplasmic progesterone receptor in hen pituitary and hypothalamic tissues. Endocrinology 102:1207–1213

Kawashima M, Kamiyoshi M, Tanaka K (1979) Nuclear progesterone receptor in the hen pituitary and hypothalamus. Endocrinol Jpn 26:501–506

Kikuyama S (1969) Alteration by neonatal hypothyroidism of the critical period for the induction of persistent estrus in the rat. Endocrinol Jpn 16:269–273

Kincl F, Maqueo A (1965) Prevention by progesterone of steroid-induced sterility in neonatal male and female rats. Endocrinology 77:859–862

Knobil E (1974) On the control of gonadotropin secretion in the rhesus monkey. Recent Prog Horm Res 30:1–36

Komisaruk BR (1967) Effects of local brain implants of progesterone in reproductive behavior in ring doves. J Comp Physiol Psychol 64:219–224

Korach KS, Muldoon TG (1974) Studies on the nature of the hypothalamic estradiol-concentrating mechanism in the male and female rat. Endocrinology 94:785–793

Kow LM, Pfaff D (1977) Induction of lordosis in female rats. Two modes of E_2 action and the effect of ADX. Horm Behav 6:259–276

Krey L, Butler W, Knobil E (1975) Surgical disconnection of the medial basal hypothalamus and pituitary function in the rhesus monkey. I. Gonadotropin secretions. Endocrinology 96:1073–1087

Krey LC, MacLusky NJ, Davies PG, Lieberburg I, Roy EJ (1982) Different intracellular mechanism underlie testosterone's suppression of basal and stimulation of cyclic luteinizing hormone release in male and female rats. Endocrinology 110:2159–2169

Kubli-Garfias C, Whalen RE (1977) Induction of lordosis behavior in female rats by intravenous administration of progestins. Horm Behav 9:380–386

Kusama T, Mabuchi T (1970) Stereotaxic atlas of the brain of *Macaca fuscata*. University of Tokyo Press, Tokyo

Laumas KR, Farooq A (1966) The uptake in vivo of [1,2-^{3}H]-progesterone by the brain and genital tract of the rat. J Endocrinol 36:95–96

Leavitt WW, Chen TJ, Allen TC, Johnston JO (1977) Regulation of progesterone receptor formation by estrogen action. Ann NY Acad Sci 286:210–225

Lee H, Davies IJ, Ryan KJ (1979) Progesterone receptor in the hypothalamic cytosol of female rats. Endocrinology 104:791–800

Lieberburg I, Krey LC, McEwen BS (1979) Sex differences in serum testosterone and in exchangeable brain cell nuclear estradiol during the neonatal period in rats. Brain Res 178:207–212

Lisk RD (1960a) A comparison of the effectiveness of intravanous, as opposed to subcutaneous, injections of progesterone for the induction of estrous behavior in the rat. Can J Biochem Physiol 38:1381–1383

Lisk RD (1960b) Estrogen-sensitive centers in the hypothalamus of the rat. J Exp Zool 145:197–207

Lisk RD, Reuter LA (1977) In vivo progesterone treatment enhances [^{3}H]estradiol retention by neural tissue of the female rat. Endocrinology 100:1652–1658

Lowry OH, Rosebrough NJ, Farr AL, Randall RJ (1951) Protein measurement with the Folin phenol reagent. J Biol Chem 193:265–175

MacLusky NJ, McEwen BS (1978) Oestrogen modulates progestin receptor concentrations in some brain regions but not in others. Nature 274:276–277

MacLusky NJ, McEwen BS (1980a) Progestin receptors in rat brain: Distribution and properties of cytoplasmic progestin binding sites. Endocrinology 106:192–202

MacLusky NJ, McEwen BS (1980b) Progestin receptors in the developing rat brain and pituitary. Brain Res 189:262–268

MacLusky NJ, Naftolin F (1981) Sexual differentiation of the central nervous system. Science 211:1294–1303

MacLusky NJ, Chaptal C, McEwen BS (1979) The development of estrogen receptor systems in the rat brain and pituitary: postnatal development. Brain Res 178:143–160

MacLusky NJ, Lieberburg I, Krey LC, McEwen BS (1980) Progestin receptors in the brain and pituitary of the bonnet monkey (*Macaca radiata*): Differences between the monkey and the rat in the distribution of progestin receptors. Endocrinology 106:185–191

Mann DR, Barraclough CA (1973) The role of estrogen and progesterone in facilitating LH release in 4-day cyclic rats. Endocrinology 93:694–699

Marrone BL, Feder HH (1977) Characteristics of ^{3}H-estrone and ^{3}H-progestin uptake and effects of progesterone on ^{3}H-estrogen uptake in brain, anterior pituitary and peripheral tissues of male and female guinea pigs. Biol Reprod 17:42–57

Martin RG, Ames BN (1961) A method for determinating the sedimentation behavior of enzyme: application to protein mixture. J Biol Chem 236:132–137

Maurer RA (1974) [^{3}H]estradiol binding macromolecules in the hypothalamus and anterior pituitary of normal female, androgenized-female and male rats. Brain Res 67:175–177

Maurer RA, Wooley DE (1974) Demonstration of nuclear [^{3}H]estradiol binding in hypothalamus and amygdala of female, androgenized-female and male rats. Neuroendocrinology 16:137–147

McEwen BS (1978 a) Sexual maturation and differentiation: the role of the gonadal steroids. Prog Brain Res 48:291–307

McEwen BS (1978 b) Gonadal steroid receptors in neuroendocrine tissues. In: Receptors and hormone Action II. O'Malley BW, Birnbaumer L (eds). Academic, London, pp 354–400

McEwen BS, Parsons B (1982) Gonadal steroid action on the brain: neurochemistry and neuropharmacology. Ann Rev Pharmacol Toxicol 22:555–598

McEwen BS, Pfaff DW (1973) Chemical and physiological approaches to neuroendocrine mechanism: attempts at integration. In: Frontiers in Neuroendocrinology. Martini L, Ganong WP (eds). Oxford University Press, New York

McEwen BS, Pfaff DW, Chaptal C, Luine V (1975) Brain cell nuclear retention of ^{3}H-estradiol doses able to promote lordosis: temporal and regional aspects. Brain Res 86:155–161

McEwen BS, de Kloet ER, Wallach G (1976) Interactions in vivo and in vitro corticoids and progesterone with cell nuclei and soluble macromolecules from rat brain regions and pituitary. Brain Res 105:129–136

McGinnis MY, Krey LC, MacLusky NJ, McEwen BS (1981 a) Characterization of steroid receptor levels in intact and ovariectomized estrogen-treated rats. An examination of the quantitative, temporal and endocrine factors which influence the neuroendocrine efficacy of an estradiol stimulus. Neuroendocrinology 33:158–161

McGinnis MY, Parsons B, Rainbow TC, Krey LC, McEwen BS (1981 b) Temporal relationship between cell nuclear progestin receptor levels and sexual receptivity following intravenous progesterone administration. Brain Res 218:365–371

Meijs-Roelofs HMA, de Greef WJ, Uilenbroek JThJ (1975) Plasma progesterone and its relationship to serum gonadotropins in immature female rats. J Endocrinol 64:329–336

Meijs-Roelofs HMA, Kramer P, Gribling-Hegger L (1981) Role of progesterone in the control of secretion of follicle-stimulating hormone in the immature female rat. J Endocrinol 89:99–106

Meyerson B (1972) Latency between intravenous injection of progestins and the appearance of estrous behavior in estrogen-treated ovariectomized rats. Horm Behav 3:1–9

Milgrom E, VuHai MT, Logeat F (1977) Use of ^{3}H-R5020 for the assay of cytosol and nuclear progesterone receptor in the rat uterus. In: Progesterone receptors in normal and neoplastic tissues. McGuire WL, Raynaud JP, Baulieu EE (eds). Raven, New York, pp 261–270

Moguilewsky M, Raynaud JP (1977) Progestin binding sites in the rat hypothalamus, pituitary and uterus. Steroids 30:99–109

Moguilewsky M, Raynaud JP (1979 a) Estrogen-sensitive progestin-binding sites in the female rat brain and pituitary. Brain Res 164:165–175

Moguilewsky M, Raynaud JP (1979 b) The relevance of hypothalamic and hypophyseal progestin receptor regulation in the induction and inhibition of sexual behavior in the female rat. Endocrinology 105:516–522

Morin LP (1977) Progesterone: inhibition of rodent sexual behavior. Physiol Behav 18:701–715

Morin F, Feder HH (1974) Hypothalamic progesterone implants and facilitation of lordosis behavior in estrogen-primed ovariectomized guinea pigs. Brain Res 70:81–93
Odell WD, Swerdloff RS (1968) Progesterone-induced luteinizing and follicle-stimulating hormone surge in postmenopausal women: a simulated ovulatory peak. Proc Natl Acad Sci USA 61:529–536
O'Malley BW, Means AR (1974) Female steroid hormones and target cell nuclei. Science 183:610–620
Onouchi T, Kato J (1983) Estrogen receptors and estrogen-inducible progestin receptors in the sexual skin of the monkey. J Steroid Biochem 18:145–151
Palkovits M (1973) Isolated removal of hypothalamic or other brain nuclei of the rat. Brain Res 59:449–450
Parsons B, MacLusky NJ, Krey LC, Pfaff DW, McEwen BS (1980) The temporal relationship between estrogen-inducible progestin receptors in the female rat brain and the time course of estrogen activation of mating behavior. Endocrinology 107:774–779
Parsons B, McGinnis MY, McEwen BS (1981) Sequential inhibition by progesterone: effects on sexual receptivity and associated changes in brain cytosol progestin binding in the female rat. Brain Res 221:149–160
Parsons B, Rainbow TC, MacLusky NJ, McEwens BS (1982) Progestin receptor levels in rat hypothalamic and limbic nuclei. J Neurosci 2:1446–1452
Pasqualini JR, Sumida C, Gelly C (1976) Cytosol and nuclear [^{3}H]-estradiol binding in the fetal tissues of guinea pig. Acta Endocrinol (Copnh) 83:811–828
Pasqualini JR, Sumida C, Nguyen BL, Gelley C (1978) Quantitative evaluation of cytosol and nuclear [^{3}H]-estradiol specific binding in the fetal brain of guinea pig during fetal ontogenesis. J Steroid Biochem 9:443:447
Pavlik EJ, Coulson PB (1976) Modulation of estrogen receptors in four different target tissues: differential effects of estrogen vs progesterone. J Steroid Biochem 7:369–376
Pfaff DW (1980) Estrogens and Brain Function: a neural analysis of a hormone-controlled mammalian reproductive behavior. Springer, New York
Pfaff DW, Kleiner M (1973) Atlas of estradiol-concentrating cells in the central nervous system of the female rat. J Comp Neurol 151:121–158
Pfaff DW, Zigmond RE (1971) Neonatal androgen effects on sexual and non-sexual behavior of adult rats tested under various hormone regimes. Neuroendocrinology 7:129–145
Pfaff DW, Gerlach JL, McEwen BS, Ferin M, Carmel P, Zimmerman EA (1976) Autoradiographic localization of hormone-concentrating cells in the brain of the female rhesus monkey. J Comp Neurol 170:279–294
Phelps CP, Sawyer CH (1969) Postnatal thyroxine modifies effects of early androgen on lordosis. Horm Behav 7:331–340
Philibert D, Raynaud JP (1973) Progesterone binding in the immature mouse and rat uterus. Steroids 22:89–98
Piva F, Karla PS, Martini L (1973) Participation of the amygdala and of the cerebellum in the feedback effects of progesterone. Neuroendocrinology 11:229–239
Plapinger L, McEwen BS, Clemens LE (1973) Ontogeny of estradiol binding sites in rat brain. II. Characteristics of a neonatal macromolecule. Endrocrinology 93:1129–1139
Powers JB (1972) Facilitation of lordosis in ovariectomized rats by intracerebral progesterone implants. Brain Res 48:311–325
Presl J, Figarova V, Herzmann J, Röhling S (1975) Changes in uptake of (^{3}H)-progesterone by female rat brain and pituitary from birth to sexual maturity. Endocrinol Exp (Bratisl) 9:51–59
Rahman SS, Billiar RB, Miguel R, Johnson W, Little B (1977) The metabolic clearance rate, the brain extraction and distribution and the uterine extraction and retention of progesterone and R5020 in estrogen-treated ovariectomized rabbits. Endocrinology 101:464–468
Rainbow TC, Davis PG, McEwen BS (1980) Anisomycin inhibits the activation of sexual behavior by estradiol and progesterone. Brain Res 194:548–555
Rainbow TC, McGinnis MY, Krey LC, McEwen BS (1982a) Nuclear progestin receptors in rat brain and pituitary. Neuroendocrinology 34:426–432

Rainbow TC, Parsons B, McEwen BS (1982b) Sex differences in rat brain oestrogen and progestin receptors. Nature 300:648–649
Rainbow TC, Parsons B, MacLusky NJ, McEwen BS (1982c) Estradiol receptor levels in rat hypothalamic and limbic nuclei. J Neurosci 2:1439–1445
Ralph CL, Fraps RM (1960) Induction of ovulation in the hen by injection of progesterone into the brain. Endocrinology 66:269–272
Ramirez VD (1973) Endocrinology of puberty. In: Handbook of physiology, Section 7, vol 2, part 1. American Physiological Society, Washington DC
Reel J, Shin Y (1975) Oestrogen-inducible uterine progesterone receptors. Characteristics in the ovariectomized immature and adult hamster. Acta Endocrinol 80:344–354
Reuter LA, Lisk RD (1976) Progesterone may act at hypothalamus and pituitary by way of enhancement of oestrogen retention. Nature 262:790–791
Roy EJ, McEwen BS (1977) An exchange assay for estrogen receptors in cell nuclei of the adult rat brain. Steroids 30:657–669
Roy EJ, MacLusky NJ, McEwen BS (1979) Antiestrogen inhibits the induction of progestin receptors by estradiol in the hypothalamus-preoptic area and pituitary. Endocrinology 104:1333–1336
Sar M, Stumpf WE (1973) Neurons of the hypothalamus concentrate [^{3}H]-progesterone or its metabolites. Science 182:1266–1268
Scatchard G (1949) The attractions of proteins for small molecules and ions. Ann NY Acad Sci 51:660–672
Schwartz SM, Blaustein JD, Wade GN (1979) Inhibition of estrous behavior by progesterone in rats: role of neural estrogen and progestin receptors. Endocrinology 105:1078–1082
Seiki K, Hattori M (1971) A more extensive study on the uptake of labelled progesterone by the hypothalamus and pituitary gland of rats. J Endocrinol 51:793–794
Seiki K, Hattori M (1973) In vivo uptake of progesterone by the hypothalamus and pituitary of the female ovariectomized rat and its relationship to cytoplasmic progesterone-binding protein. Endocrinol Jpn 20:111–119
Seiki H, Higashida M, Imanishi M, Miyamoto M, Kitagawa T, Kotani M (1968) Radioactivity in the rat hypothalamus and pituitary gland injection of labelled progesterone. J Endocrinol 41:109–110
Seiki M, Miyamoto M, Yamashita A, Kotani M (1969) Further studies on the uptake of labelled progesterone by the hypothalamus and pituitary of rats. J Endocrinol 43:129–130
Seiki K, Haruki Y, Imanishi Y, Enomoto T (1977) Further evidence of the presence of progesterone binding proteins in female rat hypothalamus. Endocrinol Jpn 24:233–238
Sheridan PJ, Sar M, Stumpf WE (1974a) Autoradiographic localization of ^{3}H-estradiol or its metabolites in the central nervous system of the developing rat. Endocrinology 94:1386–1390
Sheridan PJ, Sar M, Stumpf WE (1974b) Interaction of exogenous steroids in the developing rat brain. Endocrinology 95:1749–1753
Södersten P (1976) Lordosis behaviour in male, female and androgenized female rats. J Endocrinol 70:409–420
Spies HG, Niswender GD (1972) Effect of progesterone and estradiol on LH release and ovulation in rhesus monkeys. Endocrinology 90:257–261
Spies H, Norman R, Quadri S, Clifton D (1977) Effects of estradiol-17β on the induction of gonadotropin release by electrical stimulation of the hypothalamus in rhesus monkeys. Endocrinology 100:314–324
Stumpf WE (1970) Estrogen-neurons and estrogen-neuron systems in the periventricular brain. Am J Anat 129:207–218
Stumpf WE, Sar M (1969) Distribution of radioactivity in hippocampus and amygdala after injection of [^{3}H]-estradiol by dry-mount autoradiography. Physiologist 12:368
Stupnicka E, Massa R, Zanisi M, Martini L (1977) Role of anterior pituitary and hypothalamic metabolism of progesterone in the control of gonadotropin secretion. Prog Reprod Biol 2:88–95

Wade GN, Feder HH (1972) [1,2-^{3}H]progesterone uptake by guinea pig brain and uterus. Differential localization, time-course of uptake and metabolism, and effects of age, sex, estrogen-priming, and competing steroids. Brain Res 45:525–543

Walters MR, Clark JH (1977) Nuclear and cytoplasmic progesterone receptors in the rat uterus: Effects of R5020 and progesterone on receptor binding and subcellular compartmentalization. In: McGuire WL, Raynaud JP, Baulieu EE (eds) Progesterone receptors in normal and neoplastic tissues. Raven, New York, pp 271–286

Warembourg M (1970) Fixation de l'oestradiol-[^{3}H] au niveau des noyaux amygdaliens, septaux et du système hypothalamo-hypophysaire chez la souris femelle. Cr Hebd Séanc Acad Sci Paris 270:152–154

Warembourg M (1978 a) Radioautographic study of the brain and pituitary after (^{3}H) progesterone injection into estrogen-primed ovariectomized guinea pigs. Neurosci Lett 7:1–5

Warembourg M (1978 b) Uptake of ^{3}H labeled synthetic progestin by rat brain and pituitary. A radioautography study. Neurosci Lett 9:329–332

Warembourg M (1978 c) Radioautographic study of the rat brain, uterus and vagina after (^{3}H) R5020 injection. Mol Cell Endocrinol 12:67–79

Whalen RE, Gorzalka BB (1974) Estrogen-progesterone interactions in uterus and brain of intact and adrenectomized immature and adult rats. Endocrinology 94:214–222

Whalen RE, Luttge WG (1971 a) Differential localization of progesterone uptake in brain. Role of sex estrogen pretreatment and adrenalectomy. Brain Res 33:147–155

Whalen RE, Luttge WG (1971 b) Role of the adrenal in the preferential accumulation of progestin by mesencephalic structures. Steroids 18:141–146

Whalen RE, Massicci J (1975) Subcellular analysis of the accumulation of estrogen by the brain of male and female rats. Brain Res 89:255–264

Whalen RE, Olsen KL (1978) Chromatin binding of estradiol in the hypothalamus and cortex of male and female rats. Brain Res 152:121–131

Whalen RE, Luttge WG, Gorzalka BB (1971) Neonatal androgenization and the development of estrogen responsivity in male and female rats. Horm Behav 2:83–90

Yamaji T, Dierschke DJ, Bhattacharva AN, Knobil E (1972) The negative feedback control by estradiol and progesterone of LH secretion in the ovariectomized rhesus monkey. Endocrinology 90:771–777

Young WC (1961) The hormones and mating behavior. In: Young WC (ed) Sex internal secretions, vol II. Williams and Wilkins, Baltimore, pp 1173–1239

Effects of Estrogen and Progesterone as Revealed by Neurophysiological Methods

YASUO SAKUMA [1]

Contents

1 Introduction

The dependence of reproductive behavior on sex steroids has been well documented and reviewed for several mammalian species (Beach 1948; Baum 1979; Pfaff 1980). Since behavior in general is a final product of integrative action of the central nervous system, the example of reproductive behavior signifies that sex steroids act on the neural tissue at some level. Also compelling is the evidence of the feedback action of sex steroids and glucocorticoids, which has been shown to be mediated by limbic and hypothalamic structures (Harris 1955; Mangili et al. 1966). Steroid hormones exert their effects on the central nervous system partly through a genomic mechanism which can be blocked by inhibitors of DNA dependent RNA synthesis or protein synthesis (McEwen et al. 1979). New protein synthesis as a result of the genomic action of steroids has been clearly defined for peripheral target tissues, as, for example, in the action of estrogen in the uterus (O'Malley and Means 1974). Evidence that steroids are incorporated into central neurons includes early demonstration by autoradiography of estrogen, progesterone, and testosterone in the hypothalamus (Pfaff and Keiner 1973; Stumpf and Sar 1976; Sar and Stumpf 1973) and glucocorticoids in the hippocampus (Warembourg 1975). Biochemical studies have shown that these structures contain specific cytosolic and nuclear receptors for estrogen (Zigmond and McEwen 1970), progesterone (Kato and Onouchi 1977), and cortisol (Chytil and Toft 1972). Estrogenic induction of various enzyme activities and of structural protein synthesis, as measured by incorporation of amino acids, is attributed to this type of re-

1 Department of Physiology II, Niigata University, School of Medicine, Asahimachi 1, Niigata 951, Japan

Current Topics in Neuroendocrinology, Vol 5

ceptor mechanism. Furthermore, receptor protein itself is seemingly regulated by steroids, for instance, binding capacity for progesterone in cytosolic and nuclear fractions of the hypothalamus increases after estrogen administration (MacLusky and McEwen 1978). Interestingly, the level of receptor changes during the course of development of the brain, which has been demonstrated by autoradiography as well as by biochemical methods (Sheridan 1979; MacLusky et al. 1976). This may relate to developmental effects of estrogen or testosterone on the growth of neurites in tissue explants of the preoptic-hypothalamic complex (Toran-Allerland 1979), or on synaptogenesis in the hypothalamus or amygdala (Raisman and Field 1971, Raisman and Field 1973; Matsumoto and Arai 1980; Nishizuka and Arai 1981), either of which is known to occur during a particular "critical" period of brain development (Phoenix et al. 1959).

Another possible mode of steroid action on the central nervous system is implied by the changes in electrical activity which follow administration of the hormones. Intravenous or intraperitoneal infusion of estrogen (Yagi 1973) or glucocorticoid (Feldman and Dafny 1970; Pfaff et al. 1971) has been shown to cause changes in the rate of action potential generation in the hypothalamus and limbic structures. Electrophoretic infusion of these hormones to the vicinity of individual neurons has indicated that these effects are direct results of hormone action on specific neurons in these particular areas of the brain (Kelly et al. 1977a, b, 1978; Steiner 1971; Whitehead and Ruf 1974; Yamada 1979; Yamada and Nishida 1978). The electrical responses of central neurons to steroids are characterized by their short latency of onset, usually on the order of seconds, which is distinctively shorter than the latency for the expression of genomic effects.

The two modes of action apparently collaborate in the performance of certain autonomic, neuroendocrine, and behavioral tasks at different stages of brain development. An example of this is the estrogenic induction of the lordosis reflex, an essential element of reproductive behavior in female rodents (Pfaff 1980). Lordosis can be elicited in female rats in estrus or in ovariectomized female rats given supplements of estrogen and progesterone (Beach 1942). Progesterone, which is usually given to the animal 48–72 h after the initial estrogen priming, can be replaced by several other substances, such as a large dose of estrogen itself (Kow and Pfaff 1975), reserpine (Meyerson 1964), luteinizing hormone-releasing hormone (LH-RH) (Moss and McCann 1973; Pfaff 1973; Sakuma and Pfaff 1980a), prolactin (Harlan et al. 1983), and other peptides (Sirinathsinghji et al. 1983a, b). However, no other substance is known so far to effectively replace the priming effect of estrogen. The fact that a large dose of estrogen replaces progesterone for the behavioral activation indicates that the mechanism of estrogen action may be different in the two stages. In consideration of the importance of estrogen in the induction of lordosis, this review will focus later on possible mechanisms underlying its action on the central neurons, after a glimpse of the central nervous effect of progesterone, which heralded today's electrophysiological approach to steroid hormone action.

Attempts to examine effects of progesterone on the electrical activity of the central neurons date back to the late 1950s. Kawakami and Sawyer (1959) reported in the rabbit that progesterone alters the threshold to induce electroencephalographic (EEG) desynchronization. Single unit recording showed that large doses (400 mg) of progesterone effectively interfered with the changes in activity of the lateral hypothalamic neurons in response to vaginal probing (Barraclough and Cross 1963). However, later studies have indicated that, when given intravenously, progesterone tends to exert nonspecific anesthetizing effects on hypothalamic as well as cortical neurons (Komisaruk et al. 1967; Lincoln 1969; Ramirez et al. 1969). A dose as small as 100 μg of progesterone induced synchronization of the frontal cortical EEG, and the change was always accompanied by autonomic changes such as an increase in blood pressure (Beyer et al. 1967). Systemic administration of progesterone, via the subcutaneous route, to estrogen-primed ovariectomized female rats induces a strong lordosis response after several hours. This time course apparently coincides with an observation by Terasawa and Sawyer (1970) of an increase in hypothalamic neuronal activity induced by progesterone, in an investigation in which the EEG and blood pressure were monitored throughout to ascertain the specificity of the response. When infused intravenously, progesterone changed activities of a small number of cells in the rabbit hypothalamus within several minutes (Dufy et al. 1978). Ross et al. (1971) reported that direct implantation of crystalline progesterone into the ventral mesencephalic region resulted in a facilitation of lordosis reflex in 15 min.

Thus, although several neuroendocrine events have been pursued in an attempt to relate them to progesterone-induced electrical changes in the central neurons, we still do not know even the exact site of action of this steroid, let alone its mechanism of action through genomic or direct membrane effects. It would be appropriate to generalize that the acute effect of a large dose of progesterone seems to be a pharmacological action as an anesthetic; the mechanism of the physiological role of this steroid, which is believed to either induce lordosis through inhibition of a serotonergic mechanism (Kow et al. 1974) or terminate the period of receptivity in female rats (Morin 1977), is yet to be studied in depth.

2 Lordosis as a Model of Hormone-Dependent Behavior

Lordosis behavior is a postural reflex characterized by extreme vertebral dorsiflexion. In female rodents in estrus, this reflex can be released during the copulatory encounter by somatosensory stimuli from the male partner on the flank-tail base-perineum skin (Kow et al. 1979). Unlike the feedback action of sex steroids on gonadotropin release, in which the pituitary as well as the central neurons are involved, this behavior offers a good model to observe direct action of estrogen on the nervous system. Thus, one advantage of analyzing this behavior is that it is not necessary to worry about such a complex phenomenon as the "implantation paradox" (Bogdanove 1963) or the relative importance of the hypothalamus versus anterior pituitary, as in the feedback control system.

For the execution of lordosis, somatosensory information is not required to reach the hypothalamus. On the other hand, influence descending from the medial basal hypothalamus, which are activated by the circulating estrogen, are indispensable. Estrogen apparently releases this behavior via activation of the me-

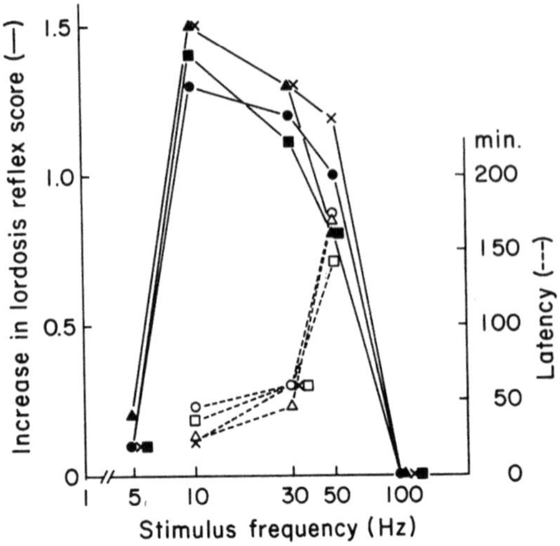

Fig. 1. Facilitation of lordosis from the ventromedial nucleus of the hypothalamus. Magnitude of increase in lordosis score (*continuous lines*) or latency to maximum response (*dashed lines*) is shown as a function of stimulus frequency. Stimulus intensity was four times threshold for each animal. Each animal received a single injection of 5 µg estradiol benzoate 4 days prior to the observation. Note that stimulation at 10–30 Hz effectively facilitated lordosis (Pfaff and Sakuma 1979 a)

dial hypothalamic neurons, since transecting all connections descending from the hypothalamus at the midbrain results in the complete loss of lordosis (Hart 1969; Kow et al. 1978). The only source of this facilitatory influence has been localized in the ventromedial nucleus of the hypothalamus. Stereotaxic implants of minute amounts of crystalline estradiol in this nucleus effectively substitute for the systemic administration (Davis et al. 1979). Electrical stimulation of the ventromedial nucleus in estrogen-primed ovariectomized female rats can facilitate lordosis, while lesion of this structure leads to loss of the behavior (Pfaff and Sakuma 1979 a, b; Fig. 1). Neurons in the ventromedial nucleus, particularly those in the ventrolateral subdivision, accumulate estradiol (Pfaff and Keiner 1973). In ovariectomized female rats, estrogen significantly increased the number of slowly firing cells in this region (Bueno and Pfaff 1976). Through these cellular changes, estrogen can cause a tonically elevated output from the ventromedial hypothalamus to the midbrain.

Another structure implicated in the central control of lordosis is the medial preoptic area, located rostral to the ventromedial hypothalamus. Although estrogen implants in this region elicit lordosis (Davis et al. 1979), the mechanism through which estrogen exerts its effect on lordosis via this structure seems to be different. For instance, lesion of this structure substantially reduces the dose of estrogen required for the induction of lordosis by the systemic route (Powers and Valenstein 1972). Electrical stimulation of the medial preoptic area disrupts lordosis in the rat and hamster (Pfaff and Sakuma 1979 a; Malsbury et al. 1980). And finally, intravenous or microelectrophoretic administration of estrogen suppresses spontaneous firing in the majority of the preoptic neurons (Dufy et al.

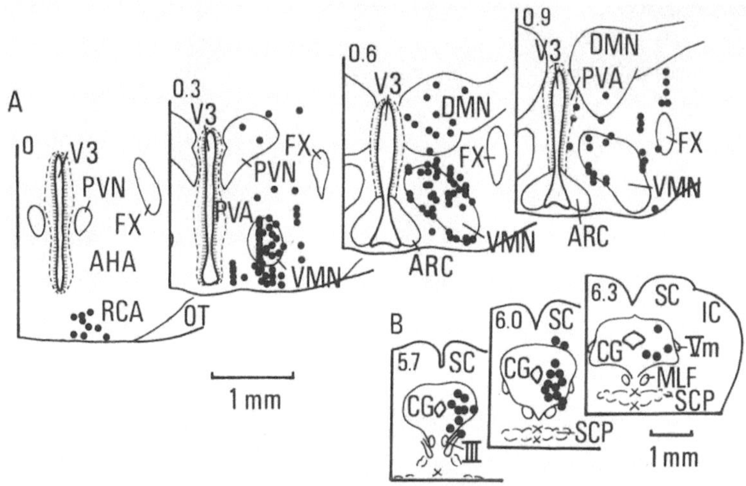

Fig. 2. A *Filled circles* indicate location of neurons in and around the ventromedial nucleus of the hypothalamus (*VMN*) which were identified by means of antidromic activation from the sites shown in **B** (*filled circles*), to have projections to the central gray matter of the mesencephalon (*CG*). *AHA*, anterior hypothalamus; *ARC*, arcuate nucleus; *DMN*, dorsomedial nucleus; *FX*, fornix; *IC*, inferior colliculus; *MLF*, medial longitudinal fascicle; *OT*, optic tract; *PVA*, periventricular area; *PVN*, paraventricular nucleus; *RCA*, retrochiasmatic area; *SC*, superior colliculus; *SCP*, superior cerebellar peduncle; *V3*, third ventricle; *III*, oculomotor nucleus; *Vm*, mesencephalic trigeminal nucleus (Sakuma and Pfaff 1982)

1976; Kelly et al. 1978). Thus, estrogen seems to act through disinhibition in the preoptic area to facilitate lordosis. It is possible that this structure serves as a relay station for several limbic outputs such as those from the septum, which is also known to suppress lordosis. It is noteworthy that estrogen can either facilitate or inhibit neuronal discharges, depending on the brain site.

Anatomical (Conrad and Pfaff 1976a; Krieger et al. 1979; Saper et al. 1976) as well as electrophysiological (MacLeod and Mayer 1980; Sakuma and Pfaff 1982) studies have shown that the mesencephalic central gray is the major target of the projections descending from the ventromedial nucleus and the medial preoptic area (Fig. 2). Electrical stimulation of the dorsal part of this structure, where a large number of terminals of ventromedial hypothalamic axons were observed (Conrad and Pfaff 1976b), facilitated the lordosis reflex in the estrogen-primed ovariectomized female rat (Sakuma and Pfaff 1979a); lesions of this area diminished the reflex (Sakuma and Pfaff 1979b). Several neurons in the central gray were antidromically identified as having axonal terminals in the gigantocellular nucleus of the medullary reticular formation (Sakuma and Pfaff 1980b).

The gigantocellular nucleus is the source of the reticulospinal tract, which has been presumed to be involved in the descending control of lordosis (Pfaff 1980). These central gray neurons exhibited a higher rate of spontaneous discharge following estrogen-supplement to ovariectomized female rats. In addition, the likelihood of antidromic action potentials propagating into the soma of the central gray neurons, which presumably corresponds to the excitability change in the

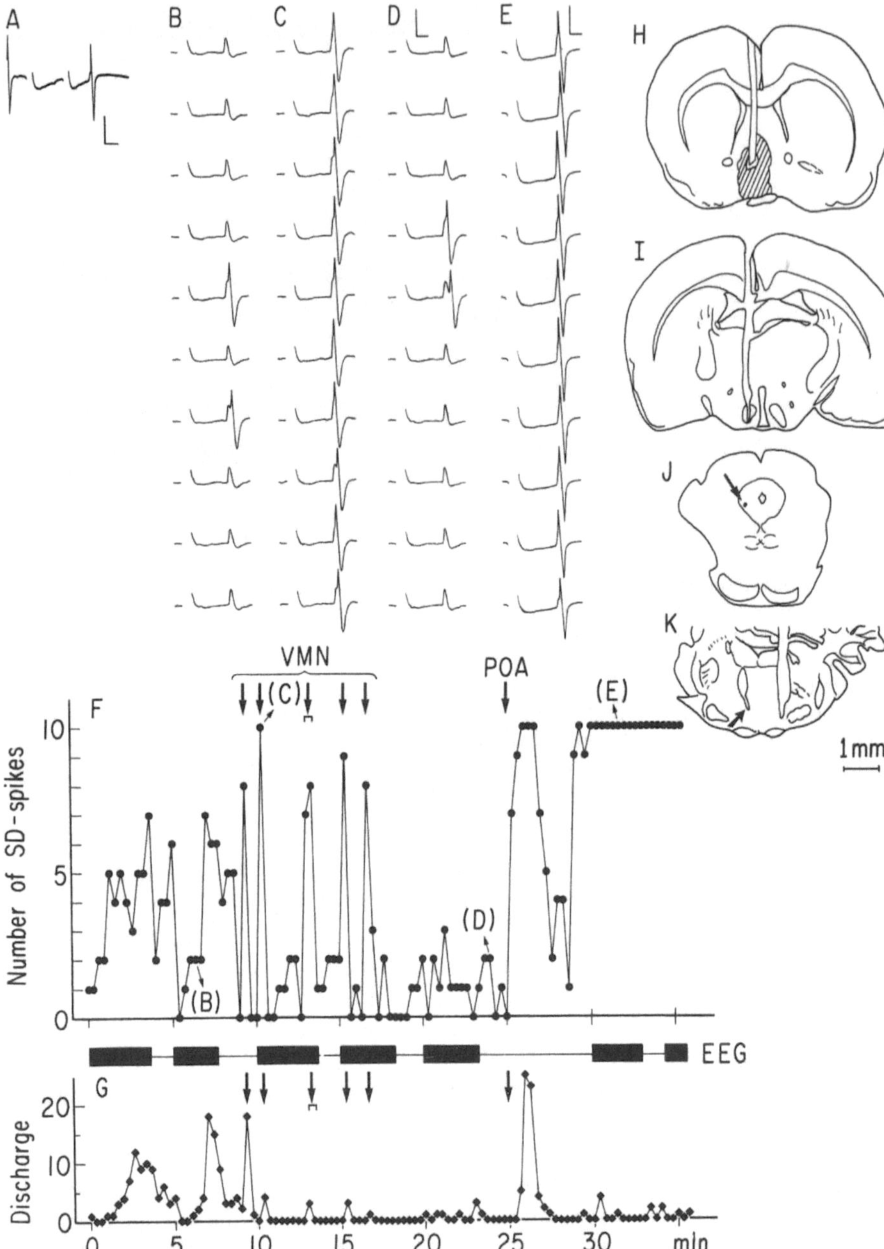

Fig. 3 A–K. Effects of ventromedial hypothalamic (*VMN*) stimulation or acute placement of lesion in the preoptic area (*POA*) on the propagation of antidromic action potential into the somatodendritic (*SD*) complex in an antidromically identified central gray cell from an ovariectomized female rat. Collision of antidromic action potential with spontaneous spike is indicated in **A. B–E** are excerpts of antidromic responses that were collected at points indicated in **F. F** shows the number of successful SD-spike formations per 20 s (10 pulses of stimuli from the gigantocellular nucleus of the medullary reticular core, **K**). The rate of resting discharge during each corresponding 20 s period appears in *G*. Continuous antidromic stimulation at 0.5 Hz had no effect on frontal cortical EEG (*broad bands* indicate periods of EEG synchronization). Stimulation of VMN and lesion of POA lesion appear in *H* (*shaded area*), and tip positions of stimulation electrodes (*arrows*) are shown in **I** and **K.** An *arrow* in **J** indicates the recording site in the lateral part of the central gray (Sakuma and Pfaff 1980c)

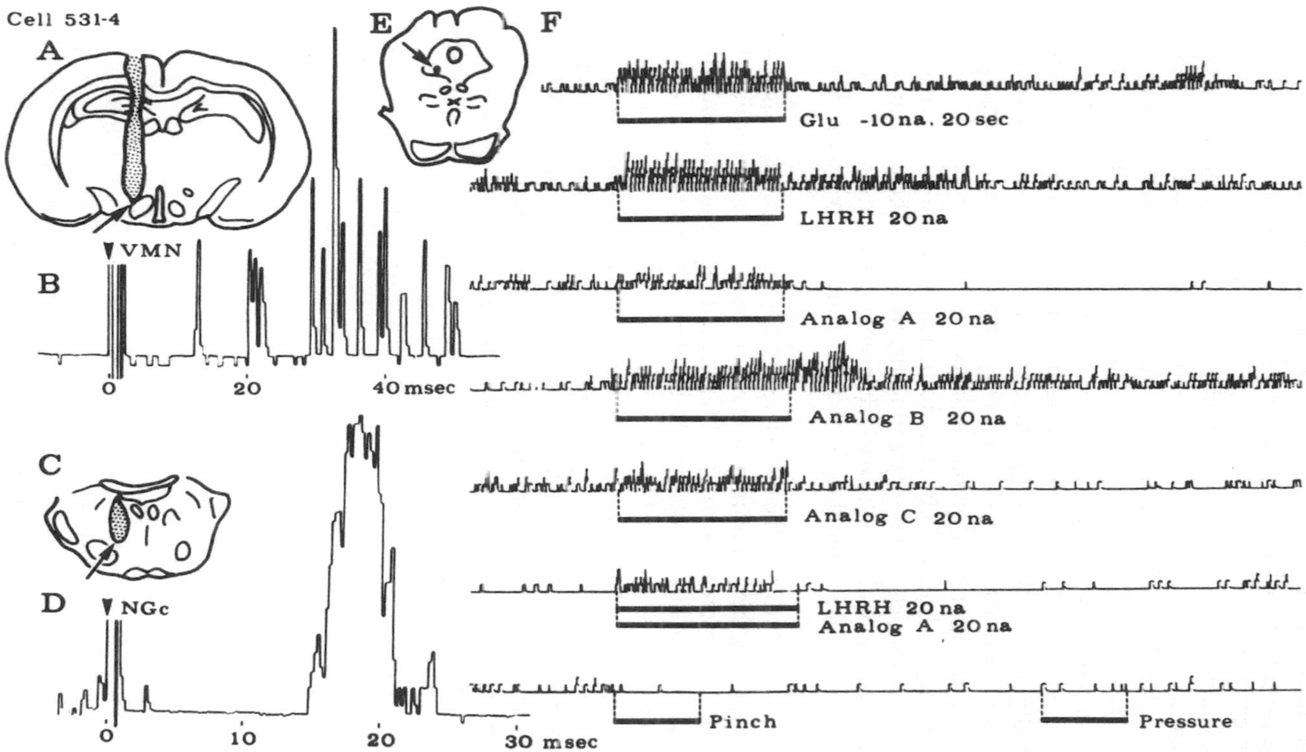

Fig. 4 A–F. Responses (**F**) of a cell in the central gray (**E,** *arrow*) to electrical stimulation of the ventromedial nucleus (*VMN*), gigantcellular nucleus of the medullary reticular core (*NGc*) and iontophoretic application of LH-RH and its analogs. Stimulation sites are indicated (*arrows*) in **A** and **C. B** and **D** are poststimulus histograms, constructed for 100 stimulation pulses. *Arrowheads* show stimulus artifacts. *Glu,* glutamate; *Analog A,* des-Gly^{10}–(D-Ala^{b})–LH-RH–EA; *Analog B,* des-Gly^{10}–LH-RH–EA; *Analog C,* des-His^{2}, des-Gly^{10}–LH-RH–EA. Note that analog A has been shown to inhibit lordosis. Tail pinch or pressure on the flank had no effect on this neuron (Akaishi and Sakuma unpublished observation)

neuronal soma caused by transsynaptic inputs, showed a marked increase following estrogen alone or stimulation of the ventromedial nucleus alone, while the estrogen-induced increase was effectively suppressed by medial preoptic stimulation (Sakuma and Pfaff 1980c). Thus, we proposed that estrogen acts on the ventromedial nucleus to facilitate the lordosis reflex, and that the effect is mediated by descending projections which impinge directly on central gray neurons and modulate their excitability (Fig. 3).

This transsynaptic modulation of excitability in the central gray neurons may depend on chemical substances such as LH-RH or prolactin. These substances facilitate lordosis when locally infused in the central gray, while antiserum against them abolishes the estrogen-induced behavior (Sakuma and Pfaff 1980a; Harlan et al. 1983). The production of these substances can be enhanced in the hypothalamus of ovariectomized female rats by estrogen supplementation (Shivers et al. 1983), supposedly through the classic genomic action of the steroid. Microiontophoresis of LH-RH and its analogs on central gray neurons induces changes in the discharge rate, in the same directions as the behavioral effects of several analogs of LH-RH (Fig. 4). Interestingly, these responses were limited to a group of central gray neurons which do not descend to the medulla. These neurons may be interneurons, corresponding to those which receive converging influences from the hypothalamus as well as from the peripheral somatosensory system (Sakuma and Pfaff 1980d).

Lordosis is a sexually dimorphic behavior. In other words, estrogen treatment is ineffective in eliciting this behavior in male rats, even after their castration. This sex difference in the responsiveness of the brain to estrogen is not determined genetically, but by exposure to testosterone during the early postnatal period of the rat. In fact, male rats castrated on the day of birth are capable of displaying lordosis when supplied with exogeneous estrogen as adults (Feder and Whalen 1965; Hart 1968); conversely, early exposure of female animals to testosterone deprives them of the ability to show lordosis (Phoenix et al. 1959). The difference between the sexes in the ability to express lordosis is considered a result of either the lack of a facilitatory mechanism for lordosis in the male, or the presence of a strong inhibition in the male (Feder 1981). The recent demonstration of the successful induction of lordosis in normal males following deafferentation of the preoptic area (Yamanouchi and Arai 1978) favors the latter theory. However, the data reviewed below illustrate that the facilitatory system may also be different between male and female rats as a result of steroid action during the early postnatal period.

3 Sexually Dimorphic Structures in the Brain

Although is has been known for years that in the human brain, functional differences between the right and left hemispheres are seemingly greater in men than in women (McGlone 1980), the first experimental evidence of sexual dimorphism of the brain was obtained at the electron-microscopic level by Raisman and Field (1971) in the synaptic organization of the rat preoptic area. They later observed that the difference depends on neonatal androgen in this species (1973). Since then, several studies have revealed structural sexual dimorphism in various areas of the central nervous system, some of which have been associated with sexually dimorphic autonomic or behavioral events. Some nuclei in the songbird archistriatum are larger in males, which sing (Nottebohm and Arnold 1976). In the rat, spinal motoneurons responsible for penile flips are prominent in males (Breedlove and Arnold 1980). Differences in the synaptic organization (Raisman and Field 1971, 1973), dendritic pattern (Greenough et al. 1977), or number of neurons (Gorski et al. 1978) in the preoptic or in the hypothalamic arcuate nucleus (Matsumoto and Arai 1980) are considered neural substrates for the different patterns in gonadotropin secretion or reproductive behavior between the sexes. In the peripheral autonomic system, the hypogastric ganglion contains a larger number of cells in prenatally hormone-treated pseudohermaphroditic mice than in the normal male or female, which may be related to the presence of the internal genitalia of both sexes in the treated animal (Suzuki et al. 1982).

The neural substrates for lordosis are another example of sexual dimorphism in the brain. Contrary to expectation, the ventromedial nucleus of the hypothalamus, which is the major facilitatory component of this female-specific behavior, is larger in male rats than in females (Matsumoto and Arai 1983). Yamanouchi and Arai (1983) assumed that inclusion of the putative inhibitory mechanism in the nucleus may contribute to this difference, on the basis of successful induction of lordosis in male rats lacking septal inputs. This interpretation requires further neurophysiological elaboration to be confirmed, since lordosis per se is a spinal segmental reflex, and, as will be discussed in detail in the next section, the descending outflow of the ventromedial nucleus to the midbrain is sexually dimorphic both in terms of morphology (wiring pattern of the axonal trajectories) and susceptivity to estrogen.

4 Functional Sex Differences of the Brain as Revealed by Electrophysiological Methods

Due to the limitations inherent in the neurophysiological technique, which deals with individual neurons rather than with the specific brain structure as an entity, attempts to elucidate the sex difference in neuronal function were not rewarding until only recently. Early electrical stimulation studies indicated that the excitability of the brain, as measured by the threshold to elicit convulsive responses, is sexually dimorphic (Wooley and Timiras 1962). Since such an effect involves numerous brain structures, it was difficult to pinpoint differences in a particular brain structure. Despite such limitations, these and other observations employing recordings of evoked potential or subcortical EEGs have provided circumstantial evidence that the neuronal function in some structures may be different between the male and female brain.

Although single unit recordings were obtained successfully from magnocellular neurosecretory cells in the hypothalamus as early as the mid-1950s, systematic application of this method to the study of elusive sexual differences in the hypothalamus and limbic structures was not attempted until 1976, when Dyer, Mac-

Leod, and Ellendorff showed that significantly more of the medial preoptic and anterior hypothalamic cells with projections to the mediobasal hypothalamus receive amygdalar inputs in male rats than in females. An earlier report (cited in Bern et al. 1973) suggested that in female rats in estrus, transsynaptic facilitation of the hypothalamic arcuate neurons from the medial preoptic area occurs at a lower threshold and a shorter latency than in males or in androgenized animals.

Only recently is decisive evidence accumulating on the functional sex differences of the brain at the neuronal level. An example is an observation on hippocampal pyramidal cells (Teyler et al. 1980). Recordings of the field potentials have shown that transsynaptic facilitation of these cells was significantly augmented by estradiol in slices from male rats, whereas testosterone had no effect. In contrast, slices from female rats in diestrus showed increased excitability to addition of testosterone, but not to estradiol. It is presumed that in this case, sex steroids exerted direct effects on the excitability of neuronal membranes. Indeed, testosterone alters the refractory period of amygdalar neuron antidromic responses elicited by medial preoptic stimulation in castrated male rats (Kendrick 1982; Kendrick and Drewett 1979; Kendrick and Drewett 1980). Parvocellular neurons in hypothalamic slices hyperpolarize with addition of estradiol to the medium (Kelly et al. 1980). Some of these neurons contained LH-RH (Kelly et al. 1982). These findings indicate that irreversible changes occur not only in the structural organization of the brain, but also at the functional level of individual neurons, as a result of brief exposure of the developing brain to sex steroids. The data reviewed below illustrate that steroid hormones exert their effects at different stages of brain development, which are seemingly relevant to several sexually dimorphic events.

5 Sex-Related Differences in Estrogen Sensitivity

Descending projections of the ventromedial hypothalamus have been shown to be different between male and female rats in terms of their connectivity, based upon the differing latencies of antidromically evoked action potentials (Sakuma and Pfaff 1981). The difference depends, at least partially, on exposure of the animal to testosterone, either secreted endogeneously or given exogeneously (Sakuma 1984; Fig. 5). Antidromic spike latency is determined by either axonal conduction velocity or conduction distance. That the present differences do not result from differences in the conduction velocity has been implied by analyses of extracellular field spread of the antidromic potential (Sakuma and Tada 1984). Thus, the conduction distance, i.e., the projection path, was suggested to be different between the sexes. Indeed, among the two major efferent systems of the ventromedial hypothalamus (Krieger et al. 1979; Saper et al. 1976), the one descending in the supraoptic commissure and through the zona incerta seemed to be dominant in females, while more of the cells decend in the periventricular system in males (Sakuma and Pfaff 1982). Disruption of the former system causes lordosis deficit more reliably than disruption of the latter (Malsbury and Daood 1978; Manogue et al. 1980). Estrogen supplementation to ovariectomized female

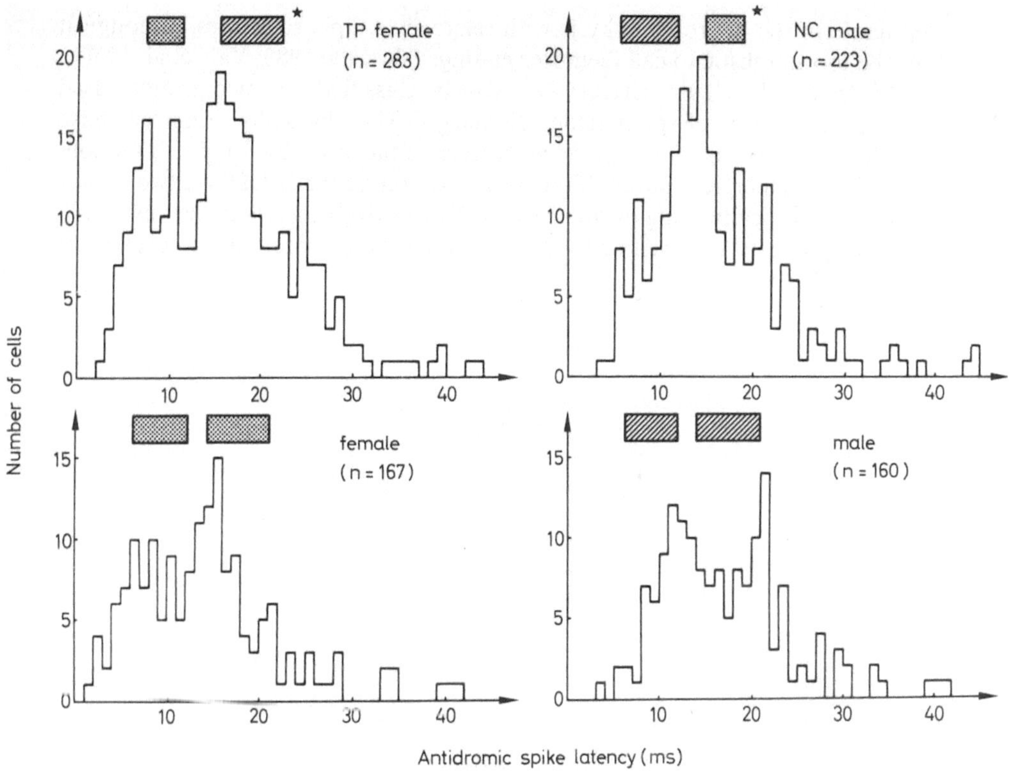

Fig. 5. Observed frequency distribution for the latency of antidromic responses of ventromedial hypothalamic neurons to electrical stimulation of the mesencephalic central gray in intact females, female rats ovariectomized as adults neonatally testosterone-treated (TP) females, neonatally castrated (*NC*) males, and intact males. *n*, number of antidromically driven cells. *Bars* above each panel indicate ranges at which neuronal population differed significantly. ($P<.05$, Kormogorov-Smirnov test) from that in the normal males (*dotted bars*) or in the ovariectomized females. *Asterisks* indicate changes caused by neonatal endocrine manipulations (Sakuma 1984)

rats has been shown to decrease the antidromic activation threshold of ventromedial axons in the central gray of the midbrain (Sakuma and Pfaff 1982). Interestingly enough, recording experiments in preparations with selective knife cuts in the efferent system revealed that this effect was seen only in those axons descending in the supraoptic commissure or through the zona incerta, and not in those in the periventricular system (Akaishi and Sakuma 1985). Changes in axonal excitability, which are indicated by a decrease in the antidromic activation threshold and a reduction in the refractory period, may occur either directly, through estrogen action on axonal membranes or indirectly, via some type of transsynaptic mechanism such as axo-axonal contacts. Since synaptic organization in the central gray is yet to be defined, it seemed impractical to explore this system further to try to explain the mechanism of estrogen action on axonal membranes. So we turned to the magnocellular neurosecretory cells in the paraventricular nucleus of the rat hypothalamus, the greater part of which are known to terminate in the neurohypophysis with relatively simple structures, although it may be too simplistic to call them free endings (Tweedle 1983; Yagi et al. 1966).

Magnocellular neurosecretory cells can be classified into two groups based upon their patterns of spontaneous discharges. The phasically firing cells have been identified as vasopressinergic, while most of the tonically firing cells are considered oxytocinergic in nature (Poulain and Wakerley 1982). Of these two groups of cells, the tonically firing oxytocinergic cells selectively increased their excitability in response to estrogen supplementation to the ovariectomized female rats

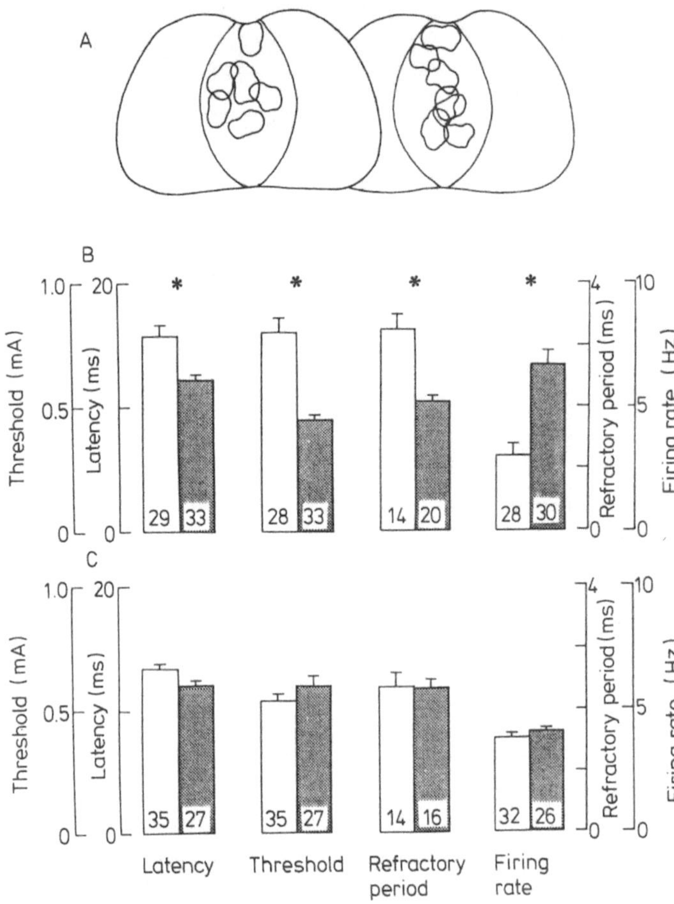

Fig. 6 A–C. Changes induced by estrogen in parameters of antidromic activation of the magnocellular neurosecretory cells in the hypothalamic paraventricular nucleus, following electrical stimulation of the neurohypophysis. **A** Locations of side-by-side stimulation electrodes are projected to the dorsal surface of the neurohypophysis. No systematic difference in the electrode placement existed between the ovariectomized (*left*) and ovariectomized and estrogen-treated (*right*) animals. Effects of estrogen were compared in antidromic spike latency, threshold for antidromic activation, relative refractory period, and firing rate, after classifying the antidromically driven cells into tonically firing (*B*) and phasically firing (*D*) groups. *Bars* are mean + SEM. *Open bars*, ovariectomized; *striped bars*, ovariectomized and estrogen-treated animals. *, $P < .001$, Mann-Whitney U-test

(Sakuma et al. 1984). In addition to all parameters of antidromic activation, the rate of spontaneous discharge also showed a significant increase (Fig. 6). It was noted that in this system, antidromic spike latency decreased in the estrogen-treated animals, along with the antidromic activation threshold and the relative refractory period. As has been reported, the increased rate of spontaneous discharge may be explained, at least partially, as a result of changes in the transsynaptic inputs from forebrain structures (Negoro and Akaishi 1981). However, other parameters of antidromic activation unanimously indicate that individual axons of the oxytocinergic cells increased their excitability to estrogen. This effect presumably operates at the level of the axonal membrane, embracing the whole axon length, otherwise the observed reduction of latency would not occur. As in the ventromedial projection to the midbrain central gray, the effect of estrogen was seen in a limited group of neurons. Only the oxytocinergic cells, but not the vasopressinergic cells, responded to estrogen treatment. It was observed that under urethane anesthesia as in the present recording experiment, plasma oxytocin but not vasopressin was significantly increased in the estrogen-treated ovariectomized female rats (Yamaguchi et al. 1979). The above observation raises a possibility that the effect may be mediated by a specific membrane mechanism. The potential existence of such a mechanism was further implied by the same kind of recording experiment in male and neonatally testosterone-treated female rats. When these animals were castrated as adults and later used in the experiment with or without estrogen administration, estrogen failed to alter all parameters of antidromic activation: a significant increase in spontaneous discharge was elicited even in these animals, but this could be a transsynaptic effect. On the other hand, in neonatally castrated male rats, oxytocinergic cells behaved much the same as in ovariectomized female rats. These observations, which were consistent with the observations in the descending projections of the ventromedial hypothalamic neurons, lead to the conclusion that some kind of estrogen receptor may exist in the axonal membrane in particular groups of neurons. The selective range of estrogen action seems to imply that the action results from a specific interaction with receptors rather than from the lipophilic nature of the steroids. The putative mechanism, which is linked with the control of the excitability of the membrane, may be irreversibly deactivated by testosterone exposure during the early postnatal period.

6 Conclusion

The concept of a genomic action of estrogen in central neurons has gained wide acceptance based upon biochemical and autoradiographic observations (McEwen et al. 1979), although exact localization of the estrogen receptor protein within subcellular fractions requires further scrutiny (Welshons et al. 1984). However, it is apparent that not all estrogen effects can be explained by this genomic action alone. Especially in many electrophysiological correlates, the effect of estrogen is obviously much more rapid than could be expected with this mode of action.

When interpreting the data obtained by means of electrophysiological techniques, one must be wary of possible mediation of estrogen effects by catecholamines. Estrogen and other steroid hormones are known to alter enzymatic activities in the metabolic pathway of catecholamine synthesis and degradation, through genomic action (McEwen et al. 1979). Some derivatives of estrogen, catechol estrogens, which are produced by aromatic hydroxylation of estrogen in the central nervous tissues (Fishman and Norton 1975), not only interfere directly with enzymatic activities (Bolt and Kappus 1976; Hiemke and Ghraf 1982) but also are presumed to interact with postsynaptic catecholamine receptors (Paden et al. 1983). These possibilities should be taken into account when considering estrogen effects on neuronal activity. Furthermore, estrogen could act directly on the presynaptic terminal to cause catecholamine release. In short-term cultures of rat hypothalamus, estrogen increased the amount of catecholamine release (Paul et al. 1979). Weissman et al. (1975) and Weissman and Johnson (1976) proposed that estrogen's effect on catecholamine release from the presynaptic terminals is mediated by the classic genomic mechanism.

Our observations on the estrogen-induced modulation of axonal excitability in the neurohypophysis refute the proposed mediation of estrogen's action by presynaptic components. A few catecholaminergic fibers, mostly dopaminergic, exist in the neurohypophysis, but they occupy a relatively restricted area adjacent to the pars intermedia (Holzbauer et al. 1978). In addition, electrophoretic studies of oxytocin indicated that this peptide induces inhibition rather than facilitation of neuronal activity (Morris et al. 1980). Thus, as far as the current data on the neurohypophyseal system and other reports on estrogen-induced electrical changes at the membrane level are concerned, this steroid seems to be acting like a modulator. For example, responses obtained in hypothalamic slices (Kelly et al. 1980) may be explained as direct effects of the steroid on the cell membrane. But we do not have any information on responses at the cell membrane level to physiological doses of estrogen, especially those demonstrated by biochemical techniques in isolated membrane preparations. In endometrial cells, estrogen has been shown to bind to the outer surfaces (Pietras and Szego 1977). It would not be unreasonable to assume that similar membrane receptors exist in neurons of the central nervous system, as in the case of classic steroid receptors.

An alternative theory is much more speculative. Protein phosphorylation is one of the most important mechanisms of cellular function (Krebs and Beavo 1979). It has been implicated in the long-term regulation of excitability of neurons (Greengard 1976; Kandel and Schwarz 1982), through activation of the sodium channel in a manner dependent on cyclic adenosine monophosphate (cyclic AMP) (Catterall 1984; Costa et al. 1982). Estrogen is known to cause accumulation of cyclic AMP in hypothalamic slices in vitro (Gunaga et al. 1974). Since estrogen enhances the efflux of catecholamines in perfusate of rat hypothalamus (Paul et al. 1979) and cyclic AMP is the intracellular second messenger for catecholamines (Siggins et al. 1971), it may be that in some cases estrogen acts presynaptically to cause the release of catecholamines, which then activate phosphorylating processes. Indeed, in the hypothalamic slices, estrogen-induced accumulation of cyclic AMP was blocked by catecholaminergic blocking agents (Gunaga et al. 1974). However, there still remains a possibility that estrogen induces a cyclic AMP-dependent protein kinase which is responsible for activation of the sodium channel without mediation of catecholamines, when we consider the excitability changes induced by estrogen in the magnocellular neurosecretory neurons.

A final question, which has been raised by several authors, is whether estrogen-induced electrophysiological changes occur solely in those cells which have been shown by autoradiography to concentrate estrogen. As has been discussed by Kelly et al. (1977b), estrogen inhibits the activity of the medial basal hypothalamic neurons in a degree similar to the percentage of neurons which concentrate estrogen. On the other hand, again in the magnocellular system, both the oxytocinergic and vasopressinergic cells have been shown to concentrate estrogen (Rhodes et al. 1982), while only the former changes electrophysiological characteristics with estrogen administration.

Thus, we do not have any concrete evidence at present on whether there is a membrane mechanism for neurotropic action of estrogen which operates independently of the classic genomic mechanism. Combined electrophysiological and morphological as well as neurochemical studies are needed to answer this formidable question.

References

Akaishi T, Sakuma Y (1985) Differential routes taken by mesencephalic projections from oestrogen-sensitive neurones in the rat ventromedial hypothalamus. J Physiol (Lond) (to be published)

Barraclough CA, Cross BA (1963) Unit activity in the hypothalamus of the cyclic female rat. Effect of genital stimuli and progesterone. J Endocrinol 26:339–359

Baum MJ (1979) Differentiation of coital behavior in mammals: a comparative analysis. Neurosci Biobehav Rev 3:265–284

Beach FA (1942) Importance of progesterone to induction of sexual receptivity in spayed female rats. Proc Soc Exp Biol Med 51:369–371

Beach FA (1948) Hormones and behavior. Hoeber, New York

Bern HA, Gorski RA, Kawashima S (1973) Long-term effects of perinatal hormone administration. Science 181:189–190

Beyer C, Ramirez VD, Whitmoyer DI, Sawyer CH (1967) Effects of hormones on the electrical activity of the brain in the rat and the rabbit. Exp Neurol 18:313–326

Bogdanove EM (1963) Direct gonad-pituitary feedback: An analysis of effects of intracranial estrogenic depots on gonadotropin secretion. Endocrinology 73:696–712

Bolt HM, Kappus H (1976) Interaction by 2-hydroxyestrogens with enzymes of drug metabolism. J Steroid Biochem 7:311–313

Breedlove SM, Arnold AP (1980) Hormone accumulation in a sexually dimorphic motor nucleus of the rat spinal cord. Science 210:564–566

Bueno J, Pfaff DW (1976) Single unit recording in hypothalamus and preoptic area of estrogen-treated and untreated ovariectomized female rats. Brain Res 101:67–78

Catterall WA (1984) The molecular basis of neuronal excitability. Science 223:653–661

Chytil F, Toft D (1972) Corticoid binding component in brain. J Neurochem 19:2877–2880

Conrad LCA, Pfaff DW (1976a) Efferents from medial basal forebrain and hypothalamus in the rat. I. An autoradiographic study of the medial preoptic area. J Comp Neurol 169:185–220

Conrad LCA, Pfaff DW (1976b) Efferents from medial basal forebrain and hypothalamus in the rat. II. An autoradiographic study of the anterior hypothalamus. J Comp Neurol 169:221–262
Costa MRC, Casnellie JE, Catterall WA (1982) Selective phosphorylation of the alpha-subunit of the sodium channel by cAMP-dependent protein kinase. J Biol Chem 257:7918–7921
Davis PG, McEwen BS, Pfaff DW (1979) Localized behavioral effects of tritiated estradiol implants in the ventromedial hypothalamus of female rats. Endocrinology 104:898–903
Dufy B, Partouche C, Dufy-Barbe L, Vincent JD (1976) Effect of oestrogen on the electrical activity of hypothalamic units: correlation with gonadotrophic hormone levels. In: Endroczi E (ed) Cellular and molecular bases of neuroendocrine processes, Akademiai Kiado, Budapest, pp 303–312
Dufy B, Dufy-Barbe L, Vincent JD (1978) Effects of gonadal steroids on the electrical activity of hypothalamic neurons. In: Vincent JD, Kordon C (eds) Biologie cellulare des processes neurosecretoires hypothalamique, CNRS, Bordeaux, pp 207–220
Dyer RG, MacLeod NK, Ellendorff F (1976) Electrophysiological evidence for sexual dimorphism and synaptic convergence in the preoptic and anterior hypothalamic areas of the rat. Proc R Soc Lond B 193:421–440
Feder HH (1981) Perinatal hormones and their role in the development of sexually dimorphic behaviors. In: Adler NT (ed) Neuroendocrinology of reproduction, Plenum, New York, pp 127–157
Feder HH, Whalen RE (1965) Feminine behavior in neonatally castrated and estrogen-treated male rats. Science 147:306–307
Feldman S, Dafny N (1970) Effect of cortisol on unit activity in the hypothalamus of the rat. Exp Neurol 27:375–387
Fishman J, Norton B (1975) Catechol estrogen formation in the central nervous system of the rat. Endocrinology 96:1054–1059
Gorski RA, Gordon JH, Shryne JE, Southam AM (1978) Evidence for a morphological sex difference in the medial preoptic area of the rat brain. Brain Res 148:333–346
Greengard P (1976) Possible role for cyclic nucleotides and phosphorylated membrane proteins in postsynaptic actions of neurotransmitters. Nature 260:101–108
Greenough WT, Carter CS, Steerman C, DeVoogt TJ (1977) Sex differences in dendritic patterns in hamster preoptic area. Brain Res 126:63–72
Gunaga KD, Kawano A, Menon KMJ (1974) In vivo effect of estradiol benzoate on the accumulation of adenosine 3′5′ monophosphate in the rat hypothalamus. Neuroendocrinology 16:273–281
Harlan RE, Shivers BD, Pfaff DW (1983) Midbrain microinfusions of prolactin increase the estrogen-dependent behavior, lordosis. Science 219:1451–1453
Harris GW (1955) Neural control of pituitary gland. Arnold, London
Hart BL (1968) Neonatal castration: influence on neural organization of sexual reflexes in male rats. Science 160:1135–1136
Hart BL (1969) Gonadal hormones and sexual reflexes in the female rat. Horm Behav 1:65–71
Hiemke C, Ghraf R (1982) Effects of short-term exposure to catecholestrogens on catecholamine turnover in the preoptic-hypothalamic brain of ovariectomized rats. Brain Res 240:295–301
Holzbauer M, Sharman DF, Godden U, Mann SP, Stephens DB (1978) Observations on the function of the dopaminergic nerves innervating the pituitary gland. Neuroscience 3:1251–1262
Kandel E, Schwarz J (1982) Molecular biology of learning: modulation of transmitter release. Science 218:433–443
Kato J, Onouchi T (1977) Specific progesterone receptors in the hypothalamus and anterior hypophysis of the rat. Endocrinology 101:920–928
Kawakami M, Sawyer CH (1959) Neuroendocrine correlates of changes in brain activity thresholds by sex steroids and pituitary hormones. Endocrinology 65:652–668

Kelly MJ, Moss RL, Dudley CA, Fawcett CP (1977a) The specificity of the response of preoptic septal area neurons to estrogen: 17 α-estradiol versus 17 β-estradiol and the response of extrahypothalamic neurons. Exp Brain Res 30:43–52
Kelly MJ, Moss RL, Dudley CA (1977b) The effects of microelectrophoretically applied estrogen, cortisol, and acetylcholine on medial preoptic septal unit activity throughout the estrous cycle of the female rat. Exp Brain Res 30:53–64
Kelly MJ, Moss RL, Dudley CA (1978) The effect of ovariectomy on the responsiveness of preoptic-septal neurons to microelectrophoresed estrogen. Neuroendocrinology 25:204–211
Kelly MJ, Kuhnt U, Wuttke W (1980) Hyperpolarization of hypothalamic parvocellular neurons by 17 β-estradiol and their identification through intracellular staining with procion yellow. Exp Brain Res 40:440–447
Kelly MJ, Ronnekleiv OK, Eskay RL (1982) Immunocytochemical localization of luteinizing hormone-releasing hormone in neurons in the medial basal hypothalamus of the female rat. Exp Brain Res 48:97–106
Kendrick KM (1982) Inputs to testosterone-sensitive stria terminalis neurones in the rat brain and the effects of castration. J Physiol (Lond) 323:437–447
Kendrick KM, Drewett RF (1979) Testosterone reduces refractory period of stria terminalis neurons in the rat brain. Science 204:877–879
Kendrick KM, Drewett RF (1980) Testosterone-sensitive neurones respond to oestradiol but not to dihydrotestosterone. Nature 286:67–68
Komisaruk BR, McDonald PG, Whitmoyer DI, Sawyer CH (1967) Effects of progesterone and sensory stimulation on EEG and neuronal activity in the rat. Exp Neurol 19:494–507
Kow L-M, Pfaff DW (1975) Induction of lordosis in female rats: two modes of estrogen action and the effect of adrenalectomy. Horm Behav 6:259–276
Kow L-M, Malsbury CW, Pfaff DW (1974) Effects of progesterone on female reproductive behavior in rats: possible modes of action and role in behavioral sex differences. In: Montagna W, Sadler W (eds) Reproductive behavior. Plenum, New York, pp 179–210
Kow L-M, Grill H, Pfaff DW (1978) Elimination of lordosis in decerebrate female rats: observations from acute and chronic preparations. Physiol Behav 20:171–174
Kow L-M, Montgomery MO, Pfaff DW (1979) Triggering of lordosis reflex in female rats with somatosensory stimulation: quantitative determination of stimulus parameters. J Neurophysiol 42:195–202
Krebs EG, Beavo JA (1979) Phosphorylation-dephosphorylation of enzymes. Annu Rev Biochem 48:923–959
Krieger MS, Conrad LCA, Pfaff DW (1979) An autoradiographic study of the efferent connections of the ventromedial nucleus of the hypothalamus. J Comp Neurol 183:785–816
Lincoln DW (1969) Effects of progesterone on the electrical activity of the forebrain. J Endocrinol 45:585–596
MacLeod NK, Mayer ML (1980) Electrophysiological analysis of pathways connecting the medial preoptic area with the mesencephalic central grey matter in rats. J Physiol (Lond) 298:53–70
McEwen BS, Davis PG, Parsons B, Pfaff DW (1979) The brain as a target for steroid hormone action. Annu Rev Neurosci 2:65–112
MacLusky NJ, Chaptal C, Lieberberg I, McEwen BS (1976) Properties and subcellular interrelationships of presumptive estrogen receptor macromolecules in the brains of neonatal and prepubertal female rats. Brain Res 114:158–165
MacLusky NJ, McEwen BS (1978) Oestrogen modulates progestin receptor concentrations in some rat brain regions but not in the others. Nature 274:276–277
McGlone J (1980) Sex differences in human brain asymmetry: a critical survey. Behav Brain Sci 3:215–263
Malsbury CW, Daood JT (1978) Sexual receptivity: critical importance of supraoptic connections of the ventromedial hypothalamus. Brain Res 159:451–457

Malsbury CW, Pfaff DW, Malsbury AM (1980) Suppression of sexual receptivity in the female hamster: neuroanatomical projections from preoptic and anterior hypothalamic electrode sites. Brain Res 181:267–284

Mangili G, Motta M, Martini L (1966) Control of adrenocorticotropic hormone secretion. In: Martini L, Ganong WF (eds) Neuroendocrinology, vol 1. Academic, New York, pp 297–370

Manogue K, Kow L-M, Pfaff DW (1980) Transections affecting reproductive behavior of female rats: the role of hypothalamic output to the midbrain. Horm Behav 14:277–302

Matsumoto A, Arai Y (1980) Sexual dimorphism in wiring pattern in the hypothalamic arcuate nucleus and its modification by neonatal hormonal environment. Brain Res 190:238–242

Matsumoto A, Arai Y (1983) Sex difference in volume of the ventromedial nucleus of the hypothalamus in the rat. Endocrinol Jpn 30:277–280

Meyerson BJ (1964) Central nervous monoamines and hormone-induced estrous behaviour in the spayed rat. Acta Physiol Scand [Suppl] 63:241

Morin LP (1977) Progesterone: inhibition of rodent sexual behavior. Physiol Behav 18:701–715

Morris R, Salt TE, Sofroniew M, Hill RG (1980) Actions of microiontophoretically applied oxytocin, and immunohistochemical localization of oxytocin, vasopressin and neurophysin in the rat caudal medulla. Neurosci Lett 18:163–168

Moss RL, McCann SM (1973) Induction of mating behavior in rats by luteinizing hormone-releasing factor. Science 181:177–179

Negoro H, Akaishi T (1981) Effect of anterior or posterior deafferentation of the hypothalamus on unit activity in the paraventricular nucleus of ovariectomized female rats with or without estrogen treatment. Endocrinol Jpn 28:37–43

Nishizuka M, Arai Y (1981) Sexual dimorphism in synaptic organization in the amygdala and its dependence on neonatal hormone environment. Brain Res 212:31–38

Nottebohm F, Arnold AP (1976) Sexual dimorphism in vocal control areas of the songbird brain. Science 194:211–213

O'Malley BW, Means AR (1974) Female steroid hormones and target cell nuclei. Science 183:610–620

Paden CM, McEwen BS, Fishman J (1983) Binding of catechol estrogens to cell membranes. In: Merriam GR, Lipsett MB (ed) Catechol estrogens, Raven, New York, pp 189–201

Paul SM, Axelrod J, Saavedra JM, Skolnick P (1979) Estrogen-induced efflux of endogenous catecholamines from the hypothalamus in vitro. Brain Res 178:499–505

Pfaff DW (1973) Luteinizing hormone releasing factor (LRF) potentiates lordosis behavior in hypophysectomized ovariectomized female rats. Science 182:1148–1149

Pfaff DW (1980) Estrogens and brain function. Springer, Berlin Heidelberg New York

Pfaff DW, Keiner M (1973) Atlas of estradiol-concentrating cells in the central nervous system of the female rat. J Comp Neurol 151:121–158

Pfaff DW, Sakuma Y (1979 a) Facilitation of the lordosis reflex of female rats from the ventromedial nucleus of the hypothalamus. J Physiol (Lond) 288:189–202

Pfaff DW, Sakuma Y (1979 b) Deficit in the lordosis reflex of female rats caused by lesions in the ventromedial nucleus of the hypothalamus. J Physiol (Lond) 288:203–210

Pfaff DW, Silva MTA, Weiss JM (1971) Telemetered recording of hormone effects on hippocampal neurons. Science 172:394–395

Phoenix CH, Goy RW, Gerall AA, Young WC (1959) Organizational action of prenatally administered testosterone propionate on the tissues mediating mating behavior in the female guinea pig. Endocrinology 65:369–382

Pietras J, Szego CM (1977) Specific binding sites for oestrogen at the outer surfaces of isolated endometrial cells. Nature 265:69–72

Poulain DA, Wakerley JB (1982) Electrophysiology of hypothalamic magnocellular neurons secreting oxytocin and vasopressin. Neuroscience 7:773–808

Powers B, Valenstein ES (1972) Sexual receptivity: facilitation by medial preoptic lesions in female rats. Science 175:1003–1005

Raisman G, Field PM (1971) Sexual dimorphism in the preoptic area of the rat. Science 173:731–733
Raisman G, Field PM (1973) Sexual dimorphism in the neuropil of the preoptic area of the rat and its dependence on neonatal androgen. Brain Res 54:1–29
Ramirez V, Komisaruk BR, Whitmoyer DI, Sawyer CH (1969) Effects of hormones and vaginal stimulation on the EEG and hypothalamic units in rats. Am J Physiol 212:1376–1384
Rhodes CH, Morrell JI, Pfaff DW (1982) Estrogen-concentrating neurophysin-containing hypothalamic magnocellular neurons in the vasopressin-deficient (Brattleboro) rat: a study combining steroid autoradiography and immunocytochemistry. J Neurosci 2:1718–1724
Ross J, Claybaugh C, Clemens LG, Gorski RA (1971) Short latency induction of estrous behavior with intracerebral gonadal hormones in ovariectomized rats. Endocrinology 89:32–38
Sakuma Y (1984) Influences of neonatal gonadectomy or androgen exposure on the sexual differentiation of the rat ventromedial hypothalamus. J Physiol (Lond) 349:273–286
Sakuma Y, Pfaff DW (1979a) Facilitation of female reproductive behavior from mesencephalic central gray in the rat. Am J Physiol 237:R278–R284
Sakuma Y, Pfaff DW (1979b) Mesencephalic mechanisms for integration of female reproductive behavior in the rat. Am J Physiol 237:R285–R290
Sakuma Y, Pfaff DW (1980a) LH-RH in the mesencephalic central grey can potentiate lordosis reflex of female rats. Nature 283:566–567
Sakuma Y, Pfaff DW (1980b) Cells of origin of medullary projections in the central gray of the rat mesencephalon. J Neurophysiol 44:1002–1011
Sakuma Y, Pfaff DW (1980c) Excitability of female rat central gray cells with medullary projection: changes produced by hypothalamic stimulation and estrogen treatment. J Neurophysiol 44:1012–1023
Sakuma Y, Pfaff DW (1980d) Convergent effects of lordosis-relevant somatosensory and hypothalamic influences on central grey cells in the rat mesencephalon. Exp Neurol 70:269–281
Sakuma Y, Pfaff DW (1981) Electrophysiologic determination of projections from ventromedial hypothalamus to midbrain central gray: differences between female and male rats. Brain Res 225:184–188
Sakuma Y, Pfaff DW (1982) Properties of ventromedial hypothalamic neurons with axons to midbrain central gray. Exp Brain Res 46:292–300
Sakuma Y, Pfaff DW (1983) Modulation of the lordosis reflex of female rats by LHRH, its antiserum and analogs in the mesencephalic central gray. Neuroendocrinology 36:218–224
Sakuma Y, Tada K (1984) Evidence that two sizes of ventromedial hypothalamic neurones project to the mesencephalic central grey matter in rats. J Physiol (Lond) 49:287–297
Sakuma Y, Akaishi T, Ohtake M (1984) Effects of estrogen on the neurosecretory cells in the rat paraventricular nucleus: differential responses in the neurons with tonic or phasic activity. Neurosci Lett [Suppl] 17:S96
Saper CB, Swanson LW, Cowan WM (1976) The efferent connections of the ventromedial nucleus of the hypothalamus of the rat. J Comp Neurol 169:409–442
Sar M, Stumpf WE (1973) Autoradiographic localization of radioactivity in the rat brain after injection of 1,2-^{3}H-testosterone using dry mount autoradiography. Endocrinology 92:251–256
Sheridan PJ (1979) Estrogen binding in the neonatal cortex. Brain Res 178:201–206
Shivers BD, Harlan RE, Morrell JI, Pfaff DW (1983) Immunocytochemical localization of luteinizing hormone-releasing hormone in male and female rat brains. Neuroendocrinology 36:1–12
Siggins GR, Hoffer BJ, Bloom FE (1971) Studies on norepinephrine-containing afferents to Purkinje cells of rat cerebellum. III. Evidence for mediation of norepinephrine effects by cyclic 3′,5′-adenosine monophosphate. Brain Res 25:535–539
Sirinathsinghji DJS, Whittington PE, Audsley A, Fraser HM (1983a) β-Endorphin regulates lordosis in female rats by modulating LH-RH release. Nature 301:62–64

Sirinathsinghji DJS, Rees LH, Rivier J, Vale W (1983 b) Corticotropin-releasing factor is a potent inhibitor of sexual receptivity in the female rat. Nature 305:232–235
Steiner FA (1971) Neurotransmitter und Neuromodulatoren. Thieme, Stuttgart
Stumpf WE, Sar M (1976) Steroid hormone target sites in the brain: the differential distribution of estrogen, progestin, androgen and glucocorticoid. J Steroid Biochem 7:1163–1170
Suzuki Y, Ishii H, Furuya Y, Arai Y (1982) Developmental changes of the hypogastric ganglion associated with the differentiation of the reproductive tract in the mouse. Neurosci Lett 32:271–276
Terasawa E, Sawyer CH (1970) Diurnal variation in the effects of progesterone on multiple unit activity in the rat hypothalamus. Exp Neurol 27:359–374
Teyler TJ, Vardaris RM, Lewis D, Rawitch AB (1980) Gonadal steroid effects on excitability of hippocampal pyramidal cells. Science 209:1017–1019
Toran-Allerand DC (1976) Sex steroids and the development of the newborn mouse hypothalamus and preoptic area in vitro: implications for sexual differentiation. Brain Res 106:407–412
Tweedle CD (1983) Ultrastructural manifestations of increased hormone release in the neurohypophysis. In: Cross BA, Leng G (eds) The neurohypophysis: structure, function and control. Elsevier, Amsterdam, pp 259–272 (Progress of brain research, vol 60)
Warembourg M (1975) Radioautographic study of the rat brain after injection of 1,2-^{3}H-corticosterone. Brain Res 89:61–70
Weissman BA, Johnson DF (1976) Possible role of dopamine in diethylstilbesterol-elicited accumulation of cyclic AMP in incubated male rat hypothalamus. Neuroendocrinology 21:1–9
Weissman BA, Daly JW, Skolnik P (1975) Diethylstilbesterol-elicited accumulation of cyclic-AMP in incubated rat hypothalamus. Endocrinology 97:1559–1566
Welshons WV, Lieberman ME, Gorski J (1984) Nuclear localization of unoccupied oestrogen receptors. Nature 307:747–749
Whitehead SA, Ruf KB (1974) Response of antidromically identified preoptic neurons in the rat to neurotransmitters and to estrogen. Brain Res 79:185–198
Wooley DE, Timiras PS (1962) The gonad-brain relationship: effects of female sex hormones on electroshock convulsions in the rat. Endocrinology 70:196–209
Yagi K (1973) Changes in firing rates of single preoptic and hypothalamic units following an intravenous administration of estrogen in the castrated female rat. Brain Res 53:343–352
Yagi K, Azuma T, Matsuda K (1966) Neurosecretory cell: capable of conducting impulse in rats. Science 154:778–779
Yamada Y (1979) Effects of testosterone on unit activity in rat hypothalamus and septum. Brain Res 172:165–168
Yamada Y, Nishida E (1978) Effects of estrogen and adrenal androgen on unit activity of the rat brain. Brain Res 142:187–190
Yamaguchi K-I, Akaishi T, Negoro H (1979) Effect of estrogen treatment on plasma oxytocin and vasopressin in ovariectomized rats. Endocrinol Jpn 26:197–205
Yamanouchi K, Arai Y (1978) Lordosis behaviour in male rats: effects of deafferentation in the preoptic area and hypothalamus. J Endocrinol 76:381–382
Yamanouchi K, Arai Y (1983) Forebrain and lower brainstem participation in facilitatory and inhibitory regulation of the display of lordosis in female rats. Physiol Behav 30:155–159
Zigmond RE, McEwen BS (1970) Selective retention of oestradiol by cell nuclei in specific brain regions of the ovariectomized rat. J Neurochem 17:889–899

Progesterone Receptors in CNS Correlated with Reproductive Behavior

BRUCE PARSONS[1,2] and DONALD PFAFF

Contents

1 Introduction

During the rat estrous cycle, estradiol (E_2) and progesterone (P) synergize to activate and terminate feminine reproductive behavior and to induce ovulation. A 36–48 h rise in plasma E_2 levels is followed by a surge in both plasma P and luteinizing hormone (LH) (Smith et al. 1975), which occurs several hours prior to the appearance of reproductive behavior on the evening of proestrus. During the initial phase of P action, P activates proceptive behaviors (Beach 1942), such as soliciting and ear-wiggling, and facilitates sexual receptivity, as measured by the frequency and quality of the lordosis posture (Ball 1937; Kuehn and Beach 1963; Hardy and DeBold 1977; Hardy 1972). Ovulation occurs during the early morning hours of estrus, and reproductive behavior disappears several hours later.

The sequential application of estrogen and P to ovariectomized rodents restores feminine reproductive behavior. The experimental manipulation of exoge-

1 Department of Pharmacology, University of Pennsylvania, School of Medicine, Philadelphia, PA 19104/USA
2 Laboratory of Neurobiology and Behavior, Rockefeller University, 1230 York Avenue, New York, NY 10021/USA

Current Topics in Neuroendocrinology, Vol 5

nous E_2 and P has permitted a more precise characterization of the E_2-P synergism which underlies the activation of receptive and proceptive behavior in the female rat. The goal of such studies describing E_2 and P action on the brain is the elucidation of a causal sequence of events, beginning with the entry of E_2 and P in the brain and ending in functionally relevant alterations in the biosynthetic and electrical activity of target neurons. At present, the information for the construction of such sequences is unavailable. While we do not yet understand how a steroid hormone, acting at the genomic level via RNA and protein synthesis, can alter the electrical activity of a neuron, we do have an extensive catalog of data regarding the cellular and biochemical consequences of steroid hormone action in the brain. Some of these consequences, which are summarized in Tables 1, 2, and 3 for estrogens, and Table 6 for progestins, include effects on protein synthesis as well as on neurotransmitter metabolism.

In this review, we will attempt to summarize and synthesize that literature which bears relation, albeit tenuously, to the identification of possible key cellular events in the central nervous system which may underlie the activation of feminine reproductive behavior and the induction of gonadotropin secretion by E_2 and in particular, by P. To this end, emphasis will be placed on temporal aspects governing ovulation and the induction of mating in the female rat by these steroid hormones. E_2 and P action on the uterus and oviduct will be considered when pertinent to our understanding of possible mechanisms governing the activation of feminine reproductive behavior by ovarian steroids.

1.1 General Properties of Steroid Hormone Receptors in the Central Nervous System

Five separate classes of steroid hormone receptors have been identified in the vertebrate brain. The intracellular macromolecules which specifically bind estrogens, androgens, progestins, glucocorticoids, or mineralocorticoids are large (6–9 *S*) thermolabile proteins which contain sulfhydryl groups (see Luine and McEwen 1976). Each class of steroid hormone receptor shows high stereospecificity for the appropriate hormones, synthetic analogs, and specific antagonists (see McEwen et al. 1979).

Autoradiography and physiochemical studies which detail the neuroanatomical localization of steroid hormone receptors have revealed a separate, though sometimes overlapping, pattern of concentration for each class of steroid. For example, in the rat, the greatest number of estrogen-, androgen-, and progestin-concentrating cells in the central nervous system is found in the hypophysiotrophic area (hypothalamus, preoptic area, and septum) and amygdala, although individual structures within these regions do not necessarily concentrate all three steroids, nor each steroid to the same degree (Pfaff and Keiner 1973; Sar and Stumpf 1973; Rainbow et al. 1982b; Parsons et al. 1982b). Progestin-concentrating cells are widely distributed throughout the rat brain (MacLusky and McEwen 1978, 1980a; Parsons et al. 1982); while the pattern of glucocorticoid concentration has been shown to differ sharply from those for estrogens, androgens, or progestins (McEwen et al. 1975a; Warembourg 1975a, b). The fact that E_2 increases the number of progestin-binding sites in a number of brain regions in the rat, most markedly in the preoptic area (POA) and mediobasal hypothalamus (MBH), suggests that at least some cells in the central nervous system contain more than one class of steroid receptor (MacLusky and McEwen 1978, 1980b; Parsons et al. 1982c; Rainbow et al. 1982b).

It is important to note that the topographical pattern of estrogen, androgen, and progestin concentration sites is very similar among many vertebrate species (see Morrell et al. 1975; Morrell and Pfaff 1981). Autoradiographical studies of the brains of the rat (Pfaff and Keiner 1973), rhesus monkey (Pfaff et al. 1974; Gerlach et al. 1975), hamster (Floody and Pfaff 1974), mink (Morrell and Pfaff 1975), two species of frogs (Kelley and Pfaff 1975; Kelley et al. 1975; Morrell et al. 1975), three species of birds (Zigmond et al. 1973; Barfield et al. 1978; Arnold et al. 1976), and the sunfish (Morrell et al. 1975) have revealed basically similar neuroanatomical patterns of estrogen and androgen concentration. The sites of estrogen and/or androgen concentration common to all species include: (1) the POA, (2) the tuberal region of the hypothalamus, (3) specific limbic forebrain structures such as the amygdala and septum, and (4) specific regions of the mesencephalon deep to the tectum (Morrell and Pfaff 1981). Although the rat (MacLusky and McEwen 1978, 1980; Parsons et al. 1981a), guinea pig (Blaustein and Feder 1979), and ring dove (Balthazart et al. 1980) show a more widespread distribution of progestin receptors than the bonnet monkey (MacLusky and McEwen 1980), the greatest concentration of progestin receptors in each species is seen in the MBH. Data gathered in many of the above species indicate that common sites of estrogen, androgen, and progestin concentration are frequently implicated in the control of reproductive behavior and neuroendocrine function (see Morrell et al. 1975; Kelley and Pfaff 1975; Kelley and Pfaff 1978; Morrell and Pfaff 1981).

1.2 General Mechanism of Steroid Hormone Action

Over the past decade, progress in the field of steroid hormone action has emphasized the role of intracellular cytoplasmic receptors, which translocate hormone into the cell nucleus and alter the expression of genetic information and protein synthesis (see O'Malley and Means 1974). This classic view of steroid-receptor dynamics depicts unoccupied receptors in cytosol, which translocate to the cell nucleus when occupied by hormone. More recent studies have suggested that all estrogen receptors are concentrated in the nuclei of target cells (King and Greene 1984; Welshons et al. 1984). The presence of E_2 receptors in cytosol may represent an artifical partitioning of unoccupied nuclear receptors during extraction. Regardless of the intracellular localization of unoccupied receptors, occupied E_2 and P receptors have been shown to induce specific mRNAs and proteins in peripheral tissue. For example, an RNA which codes for an "estrogen-induced protein" appears in the uterus of the rat within 15 min of E_2 administration (Notides and Gorski 1966; DeAngelo and Gorski 1970). In chick oviduct, P has been shown to specifically control the synthesis of the egg-white, avidin (O'Malley 1967). In the central nervous system, E_2 and P synergize to activate feminine reproductive behavior, at least partially, by altering protein synthesis in the hypothalamus (see below).

In order to understand how estrogen and progestin receptors are related to changes in sexual receptivity and gonadotropin secretion, investigators have attempted to correlate steroid-receptor interactions with changes in hypothalamic and pituitary function. Three basic types of correlations have been shown, and are discussed below: those based on neuroanatomical studies of steroid receptors and behavioral mechanisms (spatial correlations); those based on temporal properties of estrogen and P action (temporal correlations); and correlations based on the fact that RNA and protein synthesis inhibitors can block E_2 and P effects on gonadotropin secretion and sexual behavior (functional correlations). Anti-estrogens also have been used to block estrogen effects on hypothalamic and pituitary function (Functional correlations) and are reviewed elsewhere (McEwen et al. 1979).

2 Estradiol

2.1. Spatial Correlations

Autoradiographic studies of the neuroanatomy of estrogen-concentrating neurons have identified precise areas of E_2 action, and have facilitated correlations with estrogen function. In the female rat, a system of limbic, medial preoptic, and medial hypothalamic cells shows the highest concentration of estrogen receptors in the central nervous system (Pfaff 1968; Stumpf 1968; Pfaff and Keiner 1973). Recently, physiochemical analyses employing microdissection of rat brain nuclei have permitted quantification of estrogen receptors in specific regions of concentration (Rainbow et al. 1982b). The greatest number of estrogen receptors (20–35 fmol mg protein) in the rat brain are located within the periventricular POA, the medial POA, and the periventricular anterior hypothalamus, particularly the rostral portion. High numbers of E_2 receptors (10–20 fmol mg protein) are found within the arcuate nucleus, the ventromedial nucleus (particularly the lateral portion), and the bed nucleus of the stria terminalis. Fewer E_2 receptors (5–10 fmol mg protein) are located within the medial amygdala, the paraventricular nucleus, the ventral premammillary nucleus, and the dorsomedial nucleus.

Many of the above areas have been implicated in the regulation of masculine and feminine reproductive behavior, gonadotropin secretion, food intake, and aggression. Two basic types of studies have provided such information: (1) those based on lesions of specific brain areas (for reviews, see Morrell et al. 1975; Malsbury et al. 1977; Barfield and Chen 1977; McEwen et al. 1979), and (2) those based on localized implants of E_2 in discrete brain regions. Lesion studies have demonstrated that regions within the hypothalamus and preoptic area (POA) contribute differentially to the control of various neuroendocrine endpoints. For example, lesions in and in close proximity to the ventromedial nucleus (VMN) abolish the expression of feminine reproductive behavior in the female rat (see Malsbury et al. 1977); whereas lesions in the anterior hypothalamus-POA abolish masculine sexual behavior. However, lesions within the VMN-arcuate region also alter normal food intake, gonadotropin secretion, and aggressive behavior (see Martini et al. 1970; Moyer 1976). Thus, it is conceivable that a specific E_2-concentrating region, or even a single steroid-concentrating cell, may participate in the control of multiple neuroendocrine end points.

It must be noted that a disruption of fibers of passage, or an ablation of non-steroid-concentrating cells, can produce a deficit in neuroendocrine function following lesion of a specific brain region. For these regions, more direct evidence regarding specific sites of estrogen action has been provided by studies which have utilized stereotaxic application of hormone. Unfortunately, the interpretations of many such studies are clouded by the types of E_2 implants employed. Perhaps the single most important variable in determining if a neuroendocrine end points is governed by a particular locus of E_2 action is the size of the cannulae utilized. Two recent studies which have employed refined techniques of E_2 implantation which minimized steroid diffusions have indicated that the primary site of estrogen action in the activation of the lordosis reflex is the VMN (Davis et al. 1979; Rubin and Barfield 1980). A similar type of implantation study has demonstrated that the primary site of estrogen action governing positive feedback is the medial POA (Goodman 1978). Goodman (1978) also observed that stereotaxic application of E_2 into the MBH did not elicit a preovulatory LH surge. Thus, the primary sites of estrogen action governing feminine reproductive behavior and preovulatory gonadotropin secretion in the rat appear to be, at least partially, anatomically distinct.

2.2 Temporal Correlations

Studies of estrogen effects in the uterus, pituitary, and brain have been pursued to determine the temporal properties of estrogen action. In the best studied of these system, the uterus, estrogen effects have been classified according to temporal appearance. "Early uterotrophic responses" are those which appear within 0–6 h of E_2 administration, while "late uterotrophic responses" do not appear until 18–72 h after the initiation of E_2 treatment (Lan and Katzenellenbogen 1976). As discussed below, this temporal classification allows for some deductions regarding the possible relationships between estrogen receptors and E_2 action.

Uterus. Three distinct biochemical changes can be observed in the uterus within 15 min of E_2 administration (Table 1). The first of these, an increase in cyclic AMP levels, is a metabolic change which can be observed as early as 15 s after E_2 administration (Szego and Davis 1967). The other two rapid changes are biosynthetic effects of E_2, the first of which is the appearance of an RNA which codes for a protein in the uterus which is induced specifically by estrogen (Notides and Gorski 1966; DeAngelo and Gorski 1970). The second is an increase in the activity of RNA polymerase II (DNA-like RNA synthesis; Glasser et al. 1972).

Within 60 min of E_2 administration, several metabolic and biosynthetic changes can be detected in uterine cells. Metabolic changes include increases in calcium influx (Pietras and Szego 1976), in the conversion of glucose to glucose-

Table 1. Biochemical effects of estradiol in uterus

Parameter	Time first observed	References
Increase in cAMP	15 s	Szego and Davies (1967)
Increase in RNA polymerase II activity	15 min	Glasser et al. (1972)
Appearance of RNA for "induced protein"	15 min	DeAngelo and Gorski (1970) Notides and Gorski (1966)
Increased calcium influx	20 min	Pietras and Szego (1976)
Synthesis of "induced protein"	40 min	Barnea and Gorski (1970)
Albumin accumulation	60 min	Kalmer and Lowenstein (1958)
Increase in RNA polymerase I activity	60 min	Gorski (1964)
Increase in de novo fatty acid synthesis	60 min	Spooner and Gorski (1972)
Increased conversion of glucose to glucose-6-phosphate	60 min	Smith and Gorski (1968)
Altered template capacity of chromatin	60 min	Barker and Warren (1966) Glasser et al. (1972) Markaverick et al. (1978)
Increase in acidic content of chromatin	60 min	Glasser et al. (1972)
Increase in rate of peptide elongation	60 min	Whelly and Barker (1974)
Increase in cGMP	90 min	Kuehl et al. (1974)
Increase in H_2O retention	120 min	Astwood (1938) Szego and Roberts (1953)
Increased electrolytes	120 min	Kalmer and Lowenstein (1958)
Increase in histamine mobilization	180 min	Shelesnyak (1959)
Increase in net protein synthesis	240 min	Hamilton (1963) Noteboom and Gorski (1963)
Increase in number of active ribosomes	240 min	Whelly and Barker (1974)
Increase in DNA synthesis	1,080 min	Kaye et al. (1972)
Increases synthesis of RNA and protein	1,080 min	Smith and Barker (1974)
Change in mitotic activity	1,440 min	Kaye et al. (1972)

6-phosphate (Smith and Gorski 1968), and in albumin (Kalmer and Lowenstein 1958). Biosynthetic changes include the appearance of "induced protein" (Barnea and Gorski 1970), and an increase in the activity of RNA polymerase I (ribosomal RNA synthesis; Gorski 1964). Laboratories have also reported an alteration in the template capacity of uterine chromatin within 1 h of E_2 administration, as assayed by measuring transcribed RNA from purified DNA and an *E. coli* polymerase in vitro (Barker and Warren 1966; Glasser et al. 1972). Markaverich and coworkers have demonstrated that such an alteration in template capacity is a result of a marked increase in the number of RNA chain initiation sites on uterine chromatin (Markaverich et al. 1978). In addition, an increase in the acidic, but not the basic, content of uterine chromatin is observed 1 h after E_2 injection (Glasser et al. 1972).

Within 4 h of systemic E_2 administration, a number of metabolic and biosynthetic changes can be detected in the uterus. Supportive effects include increases in water retention (Szego and Roberts 1953), cyclic GMP (Kuehl et al. 1974), and electrolyte concentrations (Kalmer and Lowenstein 1958), in addition to an increase in histamine mobilization (Shelesnyak 1959). Biosynthetic changes include increased synthesis of glucose-6-phosphate dehydrogenase (Smith and Barker 1974), as well as an increase in the overall rate of protein synthesis (Hamilton 1963; Noteboom and Gorski 1963). In a cell-free system containing uterine polyribosomes, E_2 has been shown to affect first the rate of peptide elongation at each active ribosome (1 h), and to increase subsequently the number of active ribosomes (marked at 4 h, maximal at 12 h; Whelly and Barker 1974). Quantitative work has revealed almost a fivefold increase in the number of active ribosomes in the rat uterus during this 4–12 h postestradiol period (Eilon and Gorski 1972).

In contrast to the early uterotrophic responses described above, are those responses which occur 19–72 h after the initiation of E_2 treatment. Some of these late uterotrophic responses are simply extensions of many of the supportive and biosynthetic changes initiated earlier, and include increased synthesis of RNA and protein (Smith and Barker 1974). Such increased biosynthesis precedes an increase in cellular hypertrophy (Hisaw 1959), a twofold increase in uterine DNA synthesis (Kaye et al. 1972), and a change in mitotic activity (Kaye et al. 1972). These latter responses can be considered to be true growth responses of the uterus.

An additional late uterotrophic response is the restoration of the uterine cytosolic estrogen receptor concentration to 1.5 times its original value within 24 h of E_2 administration (Jensen et al. 1969; Sarff and Gorski 1971). Thus, in the uterus, E_2 facilitates the synthesis of its own protein receptor. Mester and Baulieu (1975) have characterized two separates stages of E_2 receptor replenishment: (1) an initial increase of receptors between 0.5–6 h postestradiol which is insensitive to cycloheximide, and which may be attributed to a recycling of receptors from the nucleus to the cytoplasm, and (2) an additional increase of receptors between 6–12 h which does not involve recycling of nuclear receptors, and which is abolished by cycloheximide. In other words, two separate processes, one of them dependent upon protein synthesis, contribute to the restoration of uterine cytosolic estrogen receptors.

A number of investigators have sought to determine the minimal length of time that E_2 need be retained by uterine cell nuclei in order to elicit the many biochemical changes associated with true uterine growth. In a seminal series of studies, Anderson, et al. employed both E_2 and estriol (E_3) to study temporal properties of estrogen action in the uterus. Anderson and coworkers (1972) found that E_2 and E_3 were concentrated in uterine cell nuclei rapidly and at the same rate, but were retained differentially. Maximal levels of nuclear estrogen receptors were observed within 1–3 h of E_2 or E_3 injection; however, only E_2 promoted an elevation of estrogen receptors in uterine cell nuclei at 6 h poststeroid. Furthermore, although E_2 and E_3 were equally effective in stimulating early uterotrophic responses such as RNA polymerase activity and glucose metabolism, only E_2 produced an increase in uterine wet weight at 24 h. Multiple injections of E_3, which produced high nuclear concentrations of receptor-hormone complex at all times, were effective in stimulating uterine growth (Anderson et al. 1972). On the basis of these physiological findings, Clark and Peck (1976) postulated that there are two distinct types of estrogen-binging sites in uterine cell nuclei: (1) a large number (80%–90%) of nonspecific binding sites, which retain estrogen-receptor complexes transiently, and (2) a small number (2000–3000 sites per cell; 10%–20%) of specific acceptor sites, which retain estrogen-receptor complexes for prolonged periods (>6 h). These authors hypothesized that the retention of estrogen-receptor complexes by specific acceptor sites for a minimum of 6 h is a prerequisite for stimulation of the late uterotrophic events associated with true uterine growth.

Work of other investigators has supported the concept that early and late uterotrophic responses differ with respect to their temporal requirements for nuclear E_2 retention. Lan and Katzenellenbogen (1976) chemically modified E_3 in order to produce a derivative which differed in time of retention by uterine cell nuclei. Ethinyl estriol, E_2, and E_3 were shown to accumulate maximally in uterine nuclei within 1 h of systemic administration (Lan and Katzenellenbogen 1976). However, ethinyl estriol and E_2 produced a twofold elevation in uterine nuclear estrogen receptors 24 h after administration, while E_3 merely produced control levels. Although all three compounds were effective in stimulating early uterotrophic responses (increased water retention and 2-deoxyglucose phosphorylation between 2–6 h), only ethinyl estriol and E_2 were capable of stimulating late uterotrophic responses (increased rates of 2-deoxyglucose phosphorylation at 24 h and increased DNA synthesis and uterine weight at 72 h. Surprisingly, Harris and Gorski (1978) demonstrated that although multiple injections of E_3 were as effective in stimulating uterine DNA synthesis at 24 h as a single injection of E_2, not all of these injections were required. They suggested that in uterine cell nuclei estrogen was required not continuously throughout the first 12–15 h, but only at critical phases in the response cycle.

Brain. Green et al. (1970) demonstrated that a single intravenous injection of a large dose of estradiol-17-β (100 µg) can activate P-facilitated reproductive behavior in the female rat, as measured by the lordosis quotient (LQ) score. Green and coworkers reported that no activation of feminine sexual behavior was seen within 16 h of E_2 injection, but thereafter, an almost linear increase in sexual receptivity was observed, until maximal receptivity was attained at 24 h. Parsons et al. (1980) reported that implantation of Silastic capsules, which produced physiological concentrations of E_2 in plasma, increased P-facilitated sexual receptivity as early as 18 h after capsule insertion. Subsequent increases in the frequency of lordosis were observed by Parsons and coworkers 24, 36, and 48 h after the initiation of estrogen treatment. It is important to note that the latent periods reported by Green et al. (1970) and Parsons et al. (1980) were virtually identical (17 and 18 h, respectively), even though these investigators employed vastly different regimes of E_2 treatment. Such latent periods are consistent with the hypothesis that E_2 activates feminine reproductive behavior, at least in part, by altering genomic expression in hypothalamic neurons.

Table 2. Biochemical effects of estradiol in hypothalamus

Parameter	Time first observed (h)	References
I. Short term (within 48 h poststeroid)		
Increase in activities of RNA polymerase II	2–6	Kelner et al. (1980)
Facilitation of mating behavior in female rats	17–18	Green et al. (1970) Parsons et al. (1980)
Increase in cytosolic progestin receptors	18	Parsons et al. (1980)
Increase in cystine aminopeptidase	24	Heil et al. (1971)
Increase in choline acetylase	24	Luine and McEwen (1977)

Parameter	Time observed	References
II. Long term (in excess of 48 h poststeroid)		
Increase in spontaneous activity of MBH neurons	day 10	Bueno and Pfaff (1976)
Decrease in spontaneous activity of POA neurons		
Decrease in norepinephrine synthesis in AH	hour 60	Bapra et al. (1971)
Increase in reuptake of dopamine, serotonin and norepinephrine in AH	hour 60	Cardinali and Gomez (1977)
Increase in serotonin receptors	hour 72	Biegon and McEwen (1982)
Increase in muscarinic receptors	hour 72	Rainbow et al. (1980)
Increased incorporation of lysine into protein	week 5 after ovariectomy	Litteria and Thorner (1974)
Increased in LH-RH in MBH	day 2	Kalra (1976)
Increase in G6PDH, MDH, and ICDH in MBH	day 7	Luine et al. (1974, 1975, 1977)
Decrease in MAO and TH in MBH		

Abbreviations: AH, MDH, ICDH, MAO, TH

If E_2 can activate feminine reproductive behavior within 18–24 h of administration, then any change in hypothalamic neural activity or biosynthesis which is causally related to that activation must also occur within 18–24 h of E_2 treatment. Although a number of changes in hypothalamic neural activity and biosynthesis have been observed in the ovariectomized female rodent treated with estrogen, many such changes have been produced by regimes of estrogen treatment which extend over days or weeks. Hence, the time courses of these changes are unknown. Table 2 includes a summary of cellular events in the hypothalamus which have been shown to be altered by estrogen treatment in excess of 48 h. However, it must be noted that the relevance of such changes to the activation of the lordosis reflex, which occurs 18–24 h after the initiation of estrogen treatment, remains obscure.

Relatively few events in the hypothalamus of the female rat have been shown to respond to estrogen treatment within 24 h (Table 2). Kelner et al. (1980) observed increases in the activity of RNA polymerase II, which occurred in vivo within 8 h of E_2 administration. An increase in the levels of cystine aminopeptidase 24 h after an injection of ethinyl estradiol or E_2 was reported by Heil and coworkers (1971). Luine and McEwen (1977) found that estradiol benzoate (EB) produces an increase in the in vivo levels of choline acetylase in the preoptic area within 24 h of administration.

Perhaps the best example of a biosynthetic event in the hypothalamus of the female rat which can be temporally correlated with the appearance of feminine reproductive behavior is the induction of the "estrogen-inducible" progestin receptor (Moguilewsky and Raynaud 1979). Parsons and coworkers (1980) reported that both P-facilitated mating behavior and hypothalamic progestin receptors increased monotonically when Silastic capsules containing E_2 were implanted and decreased monotonically when these capsules were removed. Several features of these data should be emphasized: (1) A significant increase in P-facilitated sexual receptivity was first observed 18 h after capsule insertion; at this time, cytosolic progestin receptors in the mediobasal hypothalamus-preoptic area (MBH-POA) had increased significantly to 26% of their maximal level; (2) after 48 h of E_2 implantation, inducible progestin receptors in the MBH-POA were maximal, and P-facilitated receptivity was approaching maximal levels; (3) after removal of Silastic capsules, P-facilitated sexual receptivity was last observed at 36 h, when inducible progestin receptors in the MBH-POA had declined to 34% of their maximal levels; (4) 48 h after capsule removal, inducible progestin receptors in the MBH-POA were approaching unstimulated levels, and sexual receptivity was not observed; and (5) following capsule removal, the decay of the progestin receptor was a first-order process, with a half-life of 24 h in the MBH-POA (Fig. 1). Based on these observations, Parsons and coworkers (1980) suggested that a "threshold level" (25%–35% maximal induction) of estrogen-inducible progestin receptors in the MBH-POA was associated with the appearance and disappearance of P-facilitated sexual receptivity in the female rat. In the female guinea pig, the induction and decay of estrogen-inducible progestin receptors in the hypothalamus-POA-septum have also been shown to be correlated temporally with the induction and decay of P-facilitated sexual receptivity (Blaustein and Feder 1979).

In contrast to its effect in the uterus, E_2 does not induce the synthesis of its own receptor protein in the hypothalamus. Following a physiological or supraphysiological injection of E_2 the hypothalamic cytosolic progestin receptor concentration is restored to its original value within 24 h (Clark, MacLusky and Naftolin, personal communication).

Many investigators have attempted to characterize the minimal length of time that E_2 need be retained by hypothalamic cell nuclei in order to activate feminine reproductive behavior. Some reports have suggested that E_2 levels in the hypothalamus must be elevated significantly at the time of mating for the full expression of lordosis (Ydstebo and Sodersten 1977; Blaustein et al. 1979). For example, Blaustein and coworkers (1979) observed that the injection of an anti-estrogen, CI-628, 18 h after the administration of E_2 and concurrently with P, in-

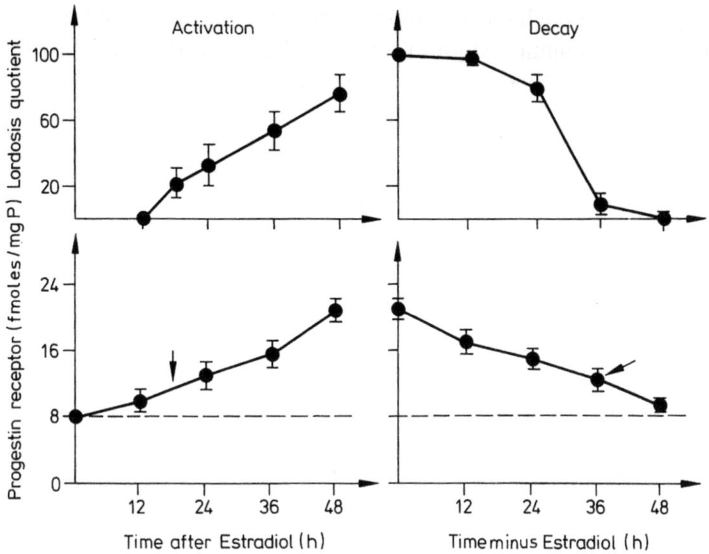

Fig. 1. The appearance and disappearance of feminine reproductive behavior and cytosolic ^{3}H-R5020 binding in the MBH-POA following E_2 treatment. Reproductive behavior was measured as the lordosis quotient (LQ), and ^{3}H-R5020 binding is expressed as femtomoles ^{3}H-R5020/mg protein. Following implantation of 5 mm E_2 Silastic capsules, LQs and ^{3}H-R5020 binding increase with time. The increase in ^{3}H-R5020 binding is significant for MBH-POA from 24 h onward ($P<0.01$; Student's t-test). Following removal of E_2 capsules, LQs and ^{3}H-R5020 binding decrease with time. *Arrows* mark the level of ^{3}H-R5020 binding at the time sexual receptivity is first and last observed. For behavioral measurements, each data point represents a minimum of 6 animals. For progestin receptor measurements, each data point represents five to seven observations (10–14 animals, pools of tissue from 2 animals)

hibited lordosis and displaced 52% of the E_2 remaining in brain cell nuclei 24 h postinjection. These investigators interpreted their data to suggest that this minute amount of E_2 (0.25% peak level, or 0.1 fmol mg protein) "appears to play a significant role in the expression of estrous behavior". However, it is conceivable that under these conditions, the mechanism of anti-estrogen displacement of receptivity did not involve simple displacement of minute amounts of steroid from hypothalamic cell nuclei.

Other reports have suggested that E_2 need not be present in hypothalamic nuclear acceptor sites during the time of induced sexual receptivity. Johnson and Davidson (1979) reported that 24 or 48 h of continuous E_2 treatment were as effective as 72 h of continuous E_2 treatment in activating the lordosis reflex at 72 h. McEwen and coworkers (1975 b) observed that extremely low (twice background, or approximately 12 cpm) levels of radioactivity were recovered from hypothalamic cell nuclei at 18 and 24 h after a behaviorally effective intravenous injection of ^{3}H-E_2. Furthermore, maximal nuclear accumulation was seen between 1 and 2 h of steroid administration. However, considerable nuclear accumulation was observed 4 h after steroid administration (30% maximal occupation in hypothalamus; McEwen et al. 1975 b). While this study strongly suggests that E_2 need not be present in the hypothalamus during the time of induced receptivity, it does not define the minimal length of time that E_2 must be retained by hypothalamic cell nuclei in order to activate feminine reproductive behavior.

It would be of great value to know the minimal period of estrogen treatment which is sufficient to activate mating behavior in the female rat. It then would be possible to describe the temporal pattern of nuclear estrogen receptor levels in the MBH-POA which is associated with the activation of lordosis. One also could begin to select those cellular events implicated in the control of sexual receptivity by their response to this regime of E_2 treatment. In a recent series of studies, Par-

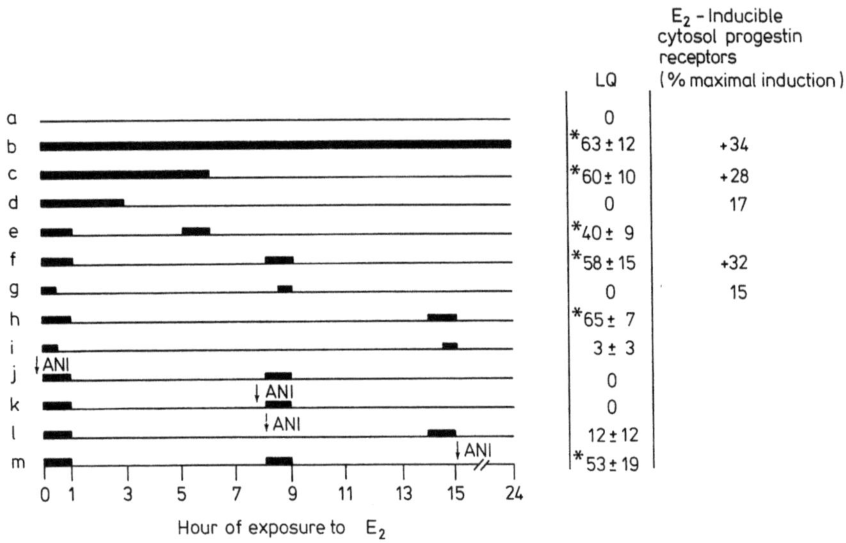

Fig. 2a–m. The effects of estradiol (*OE_2*) and anisomycin (*ANI*) on the lordosis quotient (*LQ*) and on the induction of cytosolic progestin receptors (CPR) in the MBH-POA of the female rat. Biochemical results are expressed as percent maximal induction (see text). For behavior, animals were tested with an experienced male 24 h after the initiation of E_2 treatment. Each animal received 500 µg P (s. c. in oil) 4 h before testing. In some behavioral experiments, animals received s. c. injections of the protein synthesis inhibitor, ANI, 20 mg/ml in saline. For CPR measurements, animals were killed 24 h after the insertion of Silastic capsules. No exogenous P was given before death. The temporal paradigms for these experiments are described below: **a,** No E_2, control group; **b,** E_2 0–24 h; **c,** E_2 0–6 h; **d,** E_2 0–3 h; **e,** E_2 0–1 h and 5–6 h; **f,** E_2 0–1 h and 8–9 h; **g,** E_2 0–0.5 h and 8.5–9 h; **h,** E_2 0–1 h and 14–15 h; **i,** E_2 0–0.5 h and 14.5–15 h; **j,** E_2 0–1 h and 8–9 h, ANI at −0.25 h; **k,** E_2 0–1 h and 8–9 h, ANI at 8.75 h; **l,** E_2 0–1 h and 14–15 h, ANI at 8 h; **m,** 0–1 h and 8–9 h, ANI at 15 h. CPR levels are (fmol ^{3}H-R5020/mg protein): **a,** 8.1±0.3; **b,** 12.7±0.7; **c,** 11.8±1.2; **d,** 10.3±0.7; **f,** 13.3±1.1; **g,** 10.0±0.6. CPR levels in animals which receive EB for 3 days before death (15 µg/day) are 21.1±1.2 fmol ^{3}H-R5020/mg protein

* LQ scores in groups **a–m** were compared using a one-way analysis of variance. A significant treatment effect was seen ($F=11.21$, $P<0.01$). Comparisons of individual means using the Newman-Kuels test revealed that groups **b, c, e, f,** and **h** were statistically different from the control group ($P<0.01$)

\+ CPR levels in groups **a–d, f,** and **g** were compared with each other using a one-way analysis of variance. A significant treatment was seen ($F=4.91$, $P<0.05$). Comparisons of individual means using the Newman-Kuels test revealed that groups **b, c,** and **f** were significantly different from the control group ($P<0.05$)

sons and coworkers (1981 a,, 1982 b, 1982 c) have defined such a regime of E_2 treatment – one which is just "sufficient" for activating feminine reproductive behavior – and have used this paradigm to measure associated changes in nuclear estrogen and cytosolic progestin receptors in the MBH-POA of the female rat. In short, these investigators observed that receptivity comparable to 24 h of continuous E_2 treatment was seen in animals which received only two 1-h periods of E_2 (administered using Silastic capsules), provided that the second treatment began not less than 4 nor more than 13 h after the end of the first period (*sufficient treatment;* Fig. 2). If animals received only two 0.5-h periods of E_2 during these times, receptivity was not observed (*insufficient treatment*). Moreover, *sufficient treatment* produced a threshold level of estrogen-inducible cytosolic progestin receptors in the MBH-POA (25%–35% maximal induction), whereas *insufficient treatment* did not (Fig. 2). These findings are consistent with, but not proof of, the concept that one of the neurochemical events mediated by E_2 which accompanies the activation of sexual receptivity in the female rat is the induction of cytosolic progestin receptors in the MBH-POA.

Having established that two widely separated 1-h exposures to E_2 were sufficient to activate P-facilitated receptivity, it was important to characterize the duration of estrogen retention by nuclear receptors in the MBH-POA following such brief exposure periods. E_2 treatment for 1 h elevated estrogen receptor levels in MBH-POA nuclei to 60% of capacity; whereas 0.5 h of E_2 treatment elevated nuclear estrogen receptors in the MBH-POA to approximately 30% of capacity (Parsons et al. 1981 a, b). A second 1 h period of E_2 treatment produced nuclear estrogen receptor levels in the MBH-POA which were comparable to those seen after the first 1-h treatment. After removal of the Silastic capsules, nuclear estrogen receptor levels in the MBH-POA sharply decreased in both the 1- and 0.5-h treatment groups. First-order kinetic analysis revealed that no more than 3.6% of the total nuclear estrogen receptor capacity was present in MBH-POA cells at 14 h after capsule removal (the last time at which a second 1-h period of E_2 treatment could be given to activate lordosis at 24 h) (Parsons et al. 1982 b). Thus, Parsons and coworkers concluded that an essentially discontinuous ($\leq$3.6% nuclear estrogen receptor capacity between two peaks) pattern of nuclear ^3H-E_2 retention in the MBH-POA was associated with the activation of P-facilitated lordosis, and the elevation of cytosolic progestin receptors in the MBH-POA to threshold levels. These conclusions are supported, in part, by the recent report of Sodersten et al. (1981), which indicates that two 4-h periods of E_2 treatment are adequate to activate P-facilitated lordosis at 32 h.

The kinetics of E_2 stimulation of uterine development offer parallels to some, but not all, of the above findings. It appears as if nuclear estrogen receptor levels decline more rapidly in brain cell nuclei than in uterine nuclei following E_2 administration. When nuclear estrogen receptor levels in the MBH-POA and uterus are quantified by centrifugation of homogenates through sucrose pads, they are 8% and 34% of capacity, respectively, 12 h after intravenous E_2 (Kelner et al. 1980). Kelner et al. (1980) have found that the differences in retention of nuclear E_2 receptors by the hypothalamus and the uterus were correlated with the absence of presence of "type II" receptors in cell nuclei. The presence of type II receptor in the nucleus of uterine cells is associated with the long-term uterotrophic response to E_2 (Markaverich et al. 1979). The results of Parsons and colleagues likewise suggest that nuclear estrogen receptors in the MBH-POA decline rapidly following removal of the source of E_2. However, regardless of such tissue differences in estrogen receptor subtypes, it appears that discontinuous regimes of E_2 treatment are adequate to alter physiological function in both tissues at 24 h. As discussed earlier, Harris and Gorski (1978) found that the occupation of uterine nuclear estrogen receptors at both an early and a late period was adequate to stimulate uterine DNA synthesis at 24 h. Experiments employing discontinuous E_2 stimuli suggest that there is also an essentially discontinuous requirement in brain for E_2 to activate P-facilitated reproductive behavior.

Pituitary. During the rat estrous cycle, E_2 and P synergize to induce ovulation. A 36-h rise in plasma E_2 is followed by a surge in plasma P and LH, which occurs on the afternoon of proestrus (Smith et al. 1975). Ovulation occurs during the early morning hours of estrus. Experiments of Kalra (1975) have demonstrated that the induction of preovulatory LH secretion in the rat depends upon prior exposure to estrogen, both on the day of diestrus II (0800-2300) and in the early morning of proestrus (2300 diestrus II – 0300 proestrus). Subsequent work has indicated that E_2 treatment must be initiated at least 30 h prior to "preovulatory" LH release (Krey and Parsons 1982).

If E_2 can induce a "preovulatory" LH surge from the anterior pituitary, then any changes in hypothalamic or pituitary biosynthesis which are causally related to positive feedback actions of E_2 must occur within 30–36 h of E_2 treatment. Those few cellular events in the hypothalamus which have been shown to respond to E_2 treatment within 48 h have been discussed previously. Fortunately, a number of events in the pituitary of the female rat have been shown to be affected by E_2 treatment within 36 h (Table 3); such events *may* be implicated in the induction of preovulatory gonadotropin secretion by E_2.

Several laboratories have reported that estrogen treatment alters the sensitivity of the pituitary to LH-RH. For example, Vilchez-Martinez and colleagues (1974) reported that the in vivo effect of EB on pituitary sensitivity to LH-RH was a biphasic one: decreasing sensitivity within 2–6 h, and increasing sensitivity with 14–24 h. E_2 has been shown to elicit similar effects in cells cultured from pituitary. Tang and Spies (1975) reported that incubation of cultured cells with 10^{-8} *M* E_2 decreased cell sensitivity to LH-RH at 4 h. Drouin and coworkers (1976) observed that incubation of pituitary cell cultures with 10^{-9} *M* E_2 increased the response of cultured cells to LH-RH at 10–24 h.

Estrogen has been shown to stimulate the incorporation of protein and nucleic acid precursors into pituitary LH and DNA within 36 h. There is an increase in the incorporation of ^{3}H-thymidine into pituitary DNA 30 h after in vivo administration of diethylstilbesterol (Jacobi et al. 1977). In cells cultured from pituitary, Liu and Jackson (1977) found that E_2 stimulated the in vivo incorporation of ^{3}H-alanine and ^{3}H-glucosamine into LH within 4 h of administration. Such increases could reflect either increases in precursor transport, or a stimulation of biosynthesis following estrogen treatment.

Direct effects of E_2 on pituitary biosynthesis have also been characterized temporally. Luine and McEwen (1977) found that 24 h of EB treatment increased

Table 3. Biochemical effects of estradiol in pituitary

Parameter	Time first observed (h)	References
I. Short term (within 48 h poststeroid)		
Decrease in sensitivity to LH–RH in vivo	2	Vilchez-Martinez et al. (1974)
Increased incorporation of ^{3}H-alanine and ^{3}H-glucosamine into LH	5.5	Liu and Jackson (1977)
Decrease in FSH (in culture)	6	Miller et al. (1977)
Increase in sensitivity to LH–RH (in culture)	10	Drouin et al. (1976)
Increase in cytosolic progestin receptors	12	Parsons et al. (1980)
Increase in sensitivity to LH–RH in vivo	14	Vilchez-Martinez et al. (1974)
Increased incorporation of ^{3}H-thymidine into DNA	30	Jacobi et al. (1977)

Parameter	Time first observed (day)	References
II. Long term (in excess of 48 h poststeroid)		
Increase in sensitivity to T–RH	2–4	DeLean et al. (1977)
Increase in DNA polymerase	7	Mastro and Hymer (1973)
Increase in net RNA and in RNA/DNA ratio	16	Robinson and Leavitt (1971)

pituitary levels of glucose-6-phosphate dehydrogenase, as well as total pituitary protein content, in vivo. Miller and collaborators (1977) discovered that treatment of pituitary cell cultures with 10^{-9} M E_2 for 18 h decreased the synthesis of FSH.

In order to facilitate selection of those hypothalamic and pituitary events which are causally related to the induction of preovulatory gonadotropin secretion, Krey and Parsons (1982) have defined minimal conditions of E_2 treatment which are adequate to induce P-facilitated LH release. Rats with histories of two or more consecutive 4-day estrous cycles were ovariectomized on diestrus 1 between 1900 and 2300, and were outfitted with Silastic capsules the next day ("diestrus 2"). On the presumptive day of proestrus, blood samples were collected from the tail vein at 1200, 1400, and 1600, and LH was measured by radioimmunoassay. In this paradigm, continuous E_2 stimuli which elicited an LH surge on "proestrus" had to begin by 0900 on "diestrus 2"; E_2 stimuli initiated at 1400 or 1900 on "diestrus 2" were without effect. However, a discontinuous schedule of E_2 treatment, consisting of two 3-h segments, was adequate to induce a surge of LH on "proestrus" provided that: (1) the first 3-h period was initiated at 0900 on "diestrus 2", and (2) the second 3-h period was initiated not less than 9 nor more

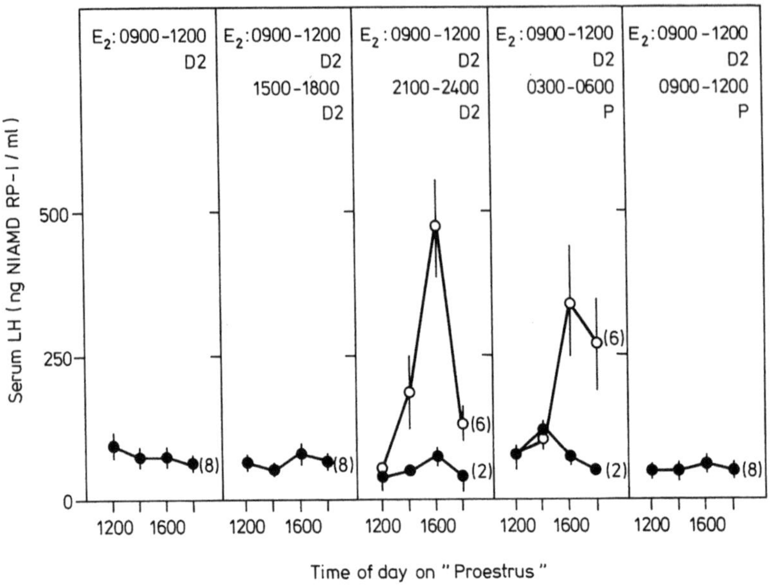

Fig. 3. Serum LH profiles on "proestrus" in female rats ovariectomized on diestrus 1 and subjected to discontinuous estrogen stimuli on "diestrus 2" (*D2*) and "proestrus" (*P*). Silastic capsules containing a 10% E_2: cholesterol mixture were implanted for the intervals presented. *Points* are means ± SEM of number of observations listed. *Open circles* signify rats responding to the estrogen stimuli with an LH surge

than 15 h after the end of the first treatment period (Fig. 3). This discontinuous schedule of E_2 treatment produced essentially discontinous nuclear estrogen receptor levels in the MBH-POA and pituitary. Surprisingly, two 1-h exposures to E_2 during these periods were not adequate to induce "preovulatory" gonadotropin secretion (Krey and Parsons, unpublished results).

What conclusions can be drawn from the experimental findings described above? First, it appears that continuous E_2 stimuli beginning at 1400 or 1900 on "diestrus 2" are ineffective because there is a latent period of 30 h between the onset of the steroid stimulus and the appearance of the gonadotropin response. Because E_2 enhancement of an event such as pituitary sensitization to GnRH does not require 30 h to develop (Vilchez-Martinez et al. 1974; Tang and Spies 1975; Krey and Parsons 1982), such an event, although perhaps necessary, is not sufficient to account for the positive feedback effects of E_2 on gonadotropin release. Perhaps the latent period for LH release is a characteristic of estrogen-sensitive neuronal mechanisms controlling GnRH secretion. It is anticipated that knowledge of an E_2 regime sufficient to induce "preovulatory" LH secretion will facilitate the identification of critical biochemical events in the hypothalamus and pituitary which are causally related to this induction.

2.3 Functional Correlations

Uterus. RNA and protein synthesis inhibitors have been used to block estrogen-mediated events in the uterus. The implication from these studies is that such events are most likely mediated by estrogen-induced changes in genomic expression. Of those events listed in Table 1, several have been shown to be blocked by actinomycin D administration, including the synthesis of "induced protein", the early increase in phospholipid and protein synthesis, and the late increases in net protein and RNA synthesis (Katzenellenbogen and Gorski 1975). It is interesting to note that none of the events listed in Table 1 have been shown to be insensitive to actinomycin D administration.

Brain and Pituitary. The genomic origin of the lordosis reflex has been implicated by the work of several investigators. Terkel et al. (1973) demonstrated that stereotaxic application of actinomycin D into the POA was effective in depressing sexual receptivity if applied before, or within 6 h of, the initiation of E_2 treatment. The extent of actinomycin D diffusion from the site of implantation was not determined in this study. Quadagno and Ho (1975) observed that stereotaxic application of cycloheximide into the POA was effective in depressing feminine reproductive behavior if given before, or within 12 h of, the initiation of estrogen treatment. Application of cycloheximide to the VMN produced an inhibition of receptivity which was comparable to that observed when cycloheximide was applied to the POA (Meinkoth et al. 1979). However, the extent of cycloheximide diffusion from the sites of implantation was not measured in either of the above studies (Quadagno and Ho 1975; Meinkoth et al. 1979).

More recent studies have employed the antibiotic protein synthesis inhibitor anisomycin (ANI) in order to provide functional correlates of E_2 action. In one such study, ANI significantly decreased the LQ score at 24 h when applied bilaterally to the VMN 6 h after the initiation of E_2 treatment (Rainbow et al. 1982a). Moreover, 4 h after ANI application, protein synthesis (as estimated by ^{3}H-leucine incorporation into soluble protein) in the MBH was inhibited by 70%, but no inhibition was observed in the POA corticomedial amygdala, or pituitary. Thus, the activation of feminine reproductive behavior by E_2 can be blocked by inhibiting protein synthesis in a discrete region of the hypothalamus consisting primarily of the VMN.

ANI has also been used to elucidate the time course of the changes in protein synthesis, following *sufficient treatment* with E_2, which are essential for the activation of the lordosis reflex in the female rat. If ANI is administered 15 min prior to the first, 15 min prior to the second, or between the two 1-h segments of E_2 treatment, little or no sexual receptivity is observed at 24 h (Parsons et al. 1981a; Parsons et al. 1982c) (Fig. 2). However, if ANI administration is delayed until 2 or 6 h after the end of the second period of E_2 treatment, no significant decrease in receptivity is observed at 24 h. Thus, at least a large portion of the protein synthesis essential for the activation of lordosis is completed within 2 h after the termination of *sufficient treatment*. Moreover, experiments which employed different doses of ANI revealed that inhibition of protein synthesis by 60% in the MBH-POA for the 1-h duration of either segment of E_2 treatment is adequate to decrease significantly, but not abolish, sexual receptivity at 24 h. To abolish sexual receptivity at 24 h, protein synthesis had to be inhibited by at least 40%–80% in the MBH-POA for 2 consecutive h, both during and 1 h after the termination of either segment of E_2 treatment (Parsons et al. 1982b).

RNA inhibitors have been used to block estrogen-induced events in the pituitary. Debeljuk et al. (1975) found that animals which had received 50 µg of E_2 18 h prior to an injection of LH-RH released more LH than animals which had not received E_2. However, if actinomycin D was given simultaneously with E_2, equivalent LH responses were observed in estrogen-treated and control groups (Debeljuk et al. 1975). This observation suggests that estrogen-induced sensitization of the pituitary to LH-RH involves a change in pituitary RNA synthesis within 18 h. However, the interpretation of all such studies is clouded by the finding that this inhibitor has effects on protein-synthesizing systems of animal cells not related to RNA synthesis (Goldstein and Penman 1973).

Several investigators have reported successes in blocking the positive and negative feedback effects of E_2 on LH release. It should be noted that E_2 may affect LH release by changing molecular events in the hypothalamus and/or the pituitary. Schally et al. (1969) found that an injection of EB (50 µg) and P (25 mg) decreased serum LH levels in the female rat 72 h later; however, actinomycin D given at h 52 abolished this decrease. Jackson (1972) found that treatment of female rats with 0.5 µg EB for 6 days, followed by a large injection of EB (50 µg) on day 7, increased serum E_2 levels within 29 h. If given simultaneously with the large EB injection, actinomycin D blocked the positive feedback effects of E_2. In a similar set of experiments, Jackson (1973) reported that cycloheximide blocked estrogen-induced increases in serum LH, when given between -5 and $+24$ h of steroid administration. These experiments suggest that both the positive and negative feedback effects of E_2 on gonadotropin secretion depend on steroid-induced changes in RNA and protein synthesis which occur in the hypothalamus and/or pituitary within 24 h of E_2 administration.

3 Progesterone

3.1 Spatial Correlations

Autoradiographic studies of the neuroanatomy of progestin-concentrating neurons which employed ^{3}H-P as a radioligand often produced negative results. One exception was the autoradiographic study by Sar and Stumpf (1973), which identified some progestin-concentrating cells in the basal hypothalamus and pituitary of the estrogen-primed guinea pig.

The availability of a high-affinity progestin, ^{3}H-R5020, facilitated the physiochemical identification of specific saturable cytosolic progestin-binding macromolecules in both ovariectomized and estrogen-stimulated rodents (Moguilewsky and Raynaud 1979; Kato and Grouch 1977; MacLusky and McEwen 1978, 1980a). Two classes of progestin binding sites have been identified in the rat (MacLusky and McEwen 1978; Kato and Grouch 1977; MacLusky and McEwen 1980b) and guinea pig (Blaustein and Feder 1979, 1980) brain. One class is distributed widely, though not uniformly, throughout the rodent brain (MacLusky and McEwen 1978, 1980b; Blaustein and Feder 1979, 1980; Munn et al. 1983) and is not induced by E_2. Although the physiological significance of the uninduced progestin receptor is unclear, it is important to note that P can elicit functional changes in the absence of estrogen – for example, P has been shown to alter EEG activity in ovariectomized rats (Arai et al. 1967).

One major feature of progestin receptors isolated from the rodent brain is the presence of a class of receptors which is induced by E_2. Physiochemical studies have demonstrated that E_2 induces progestin receptor synthesis in the MBH, POA, and pituitary of the female rat (MacLusky and McEwen 1978, 1980a) and guinea pig (Blaustein and Feder 1979, 1980). Physiochemical studies have not revealed estrogen-inducible receptors in the septal region or in the cerebellum of the rat, although these areas have been shown by physiochemical analysis to contain uninduced progestion receptors (MacLusky and McEwen 1978, 1980b, Parsons et al. 1982c). Thus, a tentative conclusion which emerges from such studies is that E_2 unduces progestin receptor synthesis primarily, but not exclusively, in those areas of the brain implicated in the control of feminine reproductive behavior and gonadotropin secretion.

Using ^{3}H-R5020 as a radioligand, Warembourg (1978) examined the neuroanatomical distribution of estrogen-inducible progestin receptors in the female rat brain with autoradiography. Few labeled cells were observed outside of the POA and MBH (Warembourg 1978). Recently, physiochemical analyses employing microdissection of rat brain nuclei have permitted quantification of both estrogen-inducible and uninduced progestin receptors in specific regions of concentration (Parsons et al. 1982c). The greatest numbers of estrogen-inducible progestin receptors (20–45 fmol mg protein) in the female rat brain are located in the periventricular POA, the medial POA, the suprachiasmatic POA, the periventricular anterior hypothalamus, the arcuate median eminence, and the VMN (Table 4). Areas which show a lesser, although significant, induction of progestin receptors (3–10 fmol mg protein) are the anterior hypothalamus, the cingulate cortex, the lateral POA, the medial nucleus of the amygdala, and the CA1 hippocampal subfield. Significant concentrations of uninduced progestin receptors were observed in many, but not all, of the above regions which contained inducible receptors, as well as in regions which did not contain inducible receptors, such as the parietal cortex and the olfactory tubercle.

Direct evidence regarding specific sites of P action in rodents has been provided by studies which have utilized stereotaxic application of hormone. Implantation of P into hypothalamic nuclei in estrogen-primed guinea pigs (Morin and Feder 1974) has been shown to facilitate proceptive and receptive behavior in this species. More specifically, intracranial application of nanogram quantities of P to the ventromedial portion of the VMN is sufficient to increase the lordosis quotient in estrogen-primed rats, while implants elsewhere in the hypothalamus are without effect (Rubin and Barfield 1984). Such findings do not necessarily preclude the participation of extrahypothalamic sites in the facilitation of feminine reproductive behavior by P.

Several studies have demonstrated that systemic administration of P to estrogen-primed males fails to potentiate lordosis (Davidson and Levine 1969;

Table 4. Progestin receptor levels in nuclei and subregions of the female rat brain

Region	Amount (fmol/mg protein ± SEM)		N
	$-E_2$	$+E_2$	
Diagonal band of Broca	2.1±1.4	1.9±1.1	4
Medial septum	2.7±1.6	0.9±0.7	4
Lateral septum	1.9±1.7	1.0±0.6	5
Olfactory tubercle	1.4±0.5[a]	1.7±0.7	4
Nucleus stria terminalis	1.4±0.9	4.6±1.2[b]	5
Medial preoptic nucleus	6.2±2.2[a]	41.8±3.5[c]	6
Suprachiasmatic preoptic nucleus	5.1±3.3	35.8±7.8[b]	5
Periventricular preoptic nucleus	1.9±1.5	48.8±8.1[c]	5
Lateral preoptic nucleus	1.4±0.7	4.9±1.5[b]	6
Caudate putamen	0.5±0.4	1.3±1.2	5
Parietal cortex	1.7±0.5[a]	2.3±0.8	5
Cingulate cortex	2.0±1.0	4.9±1.2[b]	5
Periventricular anterior hypothalamus	7.4±3.3[a]	28.9±4.9[c]	6
Anterior hypothalamus	2.2±0.8	7.5±2.1	4
Supraoptic nucleus	3.9±3.9	3.3±3.3	4
Ventromedial nucleus	4.5±1.1[a]	24.7±4.0[c]	6
Lateral ventromedial nucleus	8.9±3.4[a]	24.4±3.2[b]	4
Medial ventromedial nucleus	3.6±3.6	12.7±2.1	3
Dorsomedial nucleus	2.8±1.1[a]	5.3±0.9	5
Lateral hypothalamus	1.4±0.7	4.9±1.5[b]	6
Arcuate-median eminence	6.3±1.0[a]	40.9±5.3[c]	6
Medial amygdala	1.3±0.7	2.8±1.0[a]	5
Cortical amygdala	0.4±0.2	3.5±1.5	6
Ventral thalamus	3.2±1.7	2.3±1.3	4
Dentate gyrus	1.7±0.9	0 ±0	5
CA_1	1.0±0.6	5.1±1.1[b]	5
CA_3	0.8±0.4	0.4±0.4	5
Ventral premammillary nucleus	2.4±1.5	3.3±2.5	4
Habenula	1.5±0.9	3.0±1.8	4
Central grey	2.2±1.0	3.8±2.0	5

Cytosolic progestin receptor levels were measured according to the methods of MacLusky and McEwen (1978, 1980). Results are expressed as femtomoles of ^{3}H–R5020 per mg protein ±SEM. *N* represents number of experiments; in each experiment, tissue was pooled from three animals. Total counts per min per sample for regions with a high concentration of progestin receptors were 50–150 cpm above background. Regions with a zero to moderate level of progestin receptor ranged from 20 to 50 cpm. Nonspecific counts per min were 20–40 cpm. All samples were counted for at least 10 min

[a] Significantly different from no progestin receptor (Student's *t*-test, two-tailed test $P<0.05$

[b] Significant induction of cytosolic progestin receptors after E_2 treatment (Student's paired *t*-test, one-tailed test; $P<0.05$)

[c] Significant induction of cytosolic progestin receptors after E_2 treatment (Student's paired *t*-test, one-tailed test; $P<0.005$)

Clemens 1970; Ward 1972). Goy and McEwen (1980) suggested that sex differences in the neuroendocrine and behavioral responses to P could result from differences in the number of steroid receptors in specific locations in the brain, but classic biochemical techniques lacked sufficiently high anatomical resolution to resolve this problem. A recent study which has employed microdissection of hypothalamic nuclei to quantify progestin receptor levels in the rat brain has clarified this issue (Rainbow et al. 1982b). Although the general pattern of progestin receptor distribution is the same in the male and female rat, there are substantial quantitative differences in the receptor content of specific nuclei. There is a large sex difference in the progestin receptor content of the VMN with estrogen-treated male rats having less than 30% of the receptor levels of estrogen-treated females (Fig. 4). In fact, the level of progestin receptor observed in the male VMN after

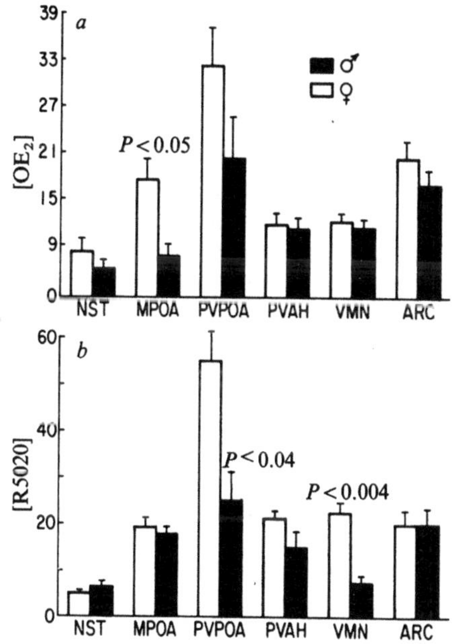

Fig. 4. Estrogen and progestin receptor levels in hypothalamic nuclei of male and female rats. Rats were gonadectomized at least 1 week and not more than 3 weeks before assays were performed. Males and females were assayed together and were matched for age and length of time after gonadectomy. Steroid receptor levels were measured in hypothalamic nuclei by the specific binding to cytosol fractions of $^3H\text{-}E_2$ (**a**) and ^{3}H-R5020, a synthetic progestin (**b**). The level of nonspecific binding corresponded to no more than 25% of the total binding. Progestin receptor measurements were made in rats given EB. Values are expressed as concentration of steroid (x 10^{-15} mol) specifically bound per mg protein: $n=9$ for paired comparisons of estrogen receptor levels in males and females; $n=6$ for paired comparisons of progestin receptor levels. Significant sex differences were seen in estrogen receptor levels in the medial POA ($P=0.045$, 2-tailed paired t comparison), in progestin receptors in the periventricular POA ($P=0.037$), and in the VMN ($P=0.003$). There were no other significant sex differences in receptor content ($P>0.120$ for all nuclei, 2-tailed paired t-test). NST, bed nucleus of the stria terminalis; MPOA, medial POA; PVPOA, periventricular POA; PVAH, periventricular portion of the anterior hypothalamus; ARC, arcuate nucleus

estrogen treatment is comparable to unstimulated levels in the female VMN. This very low level of progestin receptors in the male VMN after estrogen treatment may partially explain the inability of P to produce proceptive behaviors and increase receptivity in estrogen-primed males. The periventricular region of the preoptic area (PVPOA) also shows a significantly lower level of progestin receptors in males than in females. However, as there is little information on the possible functions of the high level of progestin receptors in the PVPOA, the significance of the sex difference in this region is unclear.

Several studies have characterized the neuroanatomical distribution of progestin receptors in species other than the rat. Although these studies have not employed autoradiography or microdissection to describe progestin receptor levels in discrete brain nuclei, they suggest that certain features of progestin receptor distribution may be common to many vertebrates. The rat (MacLusky and McEwen 1978, 1980), the guinea pig (Blaustein and Feder 1979, 1980), the ring dove (Balthazart et al. 1980), and the reptile *Anolis carolinesis* (Tobarz et al. 1981) all exhibit uninduced and estrogen-inducible progestin receptors. In most species, uninduced receptors are found throughout the brain; however, in the bonnet monkey, they appear to be restricted to the MBH and the POA (MacLusky et al. 1980). The functional significance of this finding is obscure, because the role of extrahypothalamic progestin receptors is poorly understood. The MBH is the one area in all species studied thus far which has been shown to contain estrogen-inducible progestin receptors; numerous observations suggest that these receptors play an integral role in the activation of feminine reproductive behavior (see below, and Temporal and Functional Correlations). With the exception of the bonnet monkey, all of the above species showed an increase in progestin receptor levels in the POA following estrogen treatment. Such a difference may be related to species differences in the neural control of ovulation.

3.2 Ontogeny and Hormonal Modification of Progestin Receptors During Development

Cytoplasmic progestin receptors are present in the rat brain and pituitary as early as the first few days of postnatal life (MacLusky and McEwen 1980a). Levels of uninduced receptors in the pituitary, MBH-POA, and cortex increase rapidly during the first week of life, reaching adult levels in the MBH-POA and more than 60% above adult levels in the cortex by day 10. At this time, but not before, an increase in brain and pituitary progestin receptors can be elicited by the administration of EB (Fig. 5; MacLusky and McEwen 1980). Because female rats have been reported to show lordosis behavior (Sodersten 1975; Williams 1979) and surges in gonadotropin release (Puig-Duran and MacKinnon 1978) in response to estrogen treatment between, but not before, 9–14 postnatal days, it has been suggested that the development of the estrogen-inducible progestin receptor system may contribute to the development of the lordosis reflex in the female rat (MacLusky and McEwen 1980).

The sexually dimorphic distribution of the estrogen-inducible progestin receptor system in the rat has been shown to be determined by the perinatal hormonal

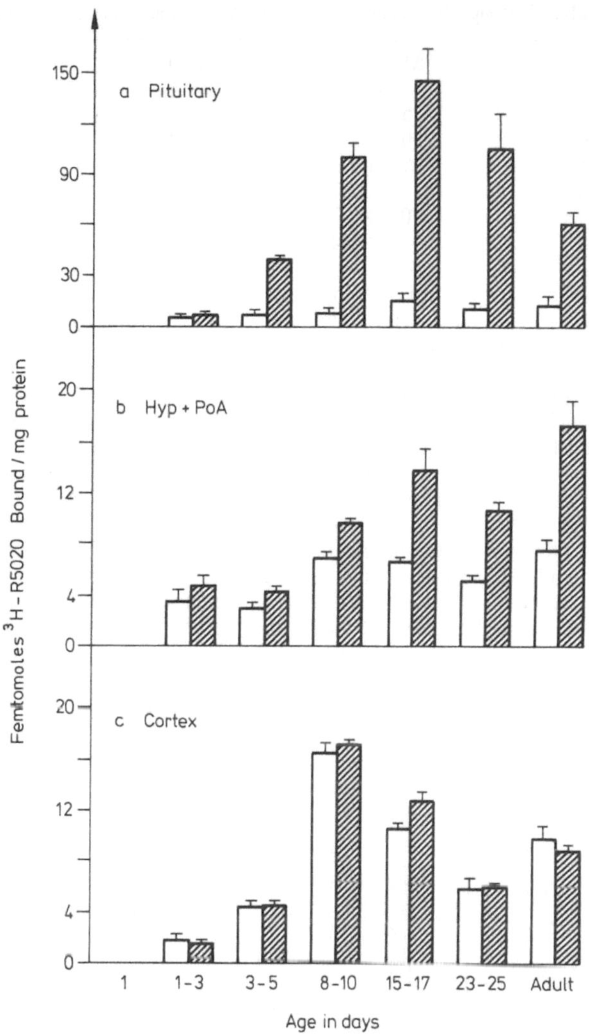

Fig. 5. Ontogeny of cytosol high-affinity ^{3}H-R5020 binding sites. Animals were injected subcutaneously 44 h before sacrifice with either R2858 (250 µg/kg; *solid bars*) in 50 µl oil, or the oil vehicle alone (*open bars*). Where an age range is given, the first figure represents the age at injection. Adult rats were ovariectomized 7 days before use. All others were used intact. Saturable ^{3}H-R5020 binding, defined as the difference between the results in the presence and absence of unlabeled R5020, is expressed per mg cytosol protein. Results represent the means ± SEM of either 4 (adults) or 3 (rest) observations. Total pituitary saturable ^{3}H-R5020 binding, expressed as femtomoles/pituitary, was as follows: *days 1–3:* oil, 0.11 ± 0.07; R2858, 0.17 ± 0.01; *days 3–5:* oil, 0.16 ± 0.03; R2858, 0.98 ± 0.07; *days 8–10:* oil, 0.44 ± 0.03; R2858, 5.04 ± 0.18; *days 15–17:* oil, 0.70 ± 0.19; R2858, 7.44 ± 1.08; *days 23–25:* oil, 1.54 ± 0.42; R2858, 18.0 ± 2.2; *adult:* oil, 16.5 ± 4.1; R2858, 97.6 ± 10.3

Table 5. Progestin receptor levels in microdissected nuclei (fmol per mg protein)

	Male	Female	ATD Male	EB female
NST	5.8+0.7	8.3+1.1	7.3+0.6	9.7+ 2.2
mPOA	20.0+2.3	27.0+3.8	22.1+2.1	22.3+ 4.1
PVPOA	29.2+3.2[c]	46.0+6.0	32.6+4.0	37.5+12.4
PVAH	17.8+2.7	23.6+7.2	17.8+2.7	30.6+ 4.6
VMN	10.1+1.2[a,c]	19.2+2.1[b]	16.6+0.5[b]	11.0+ 1.8[a]
ARC	17.7+1.4	20.7+1.7	17.1+0.3	20.3+ 3.8

NST, nucleus stria terminalis; mPOA, medial preoptic area; PVPOA, periventricular preoptic area; PVAH, periventricular anterior hypothalamus; VMN, ventromedial nucleus; ARC, arcuate nucleus

The effects of hormonal manipulation during perinatal development on the induction of progestin receptors by EB are shown. Analysis of variance tests revealed a significant treatment effect of perinatal EB and ATD administration on cytosolic progestin receptor (CPR) levels in the VMN ($F=8.23$; $P<0.001$), but not in other brain regions

[a] Comparisons of individual means in the VMN using the Newman-Keuls test revealed that CPR induction in normal females and ATD-treated males differed significantly from that in normal males and EB-treated females ($P<0.05$)

[b] There were no significant differences in VMN CPR levels between normal females and ATD-treated males, nor between normal males and EB-treated females

[c] Student's *t*-test revealed a sex difference (normal males vs normal females) in CPR levels in the VMN ($P<0.05$) and the PVPOA ($P<0.05$), but not in other brain regions

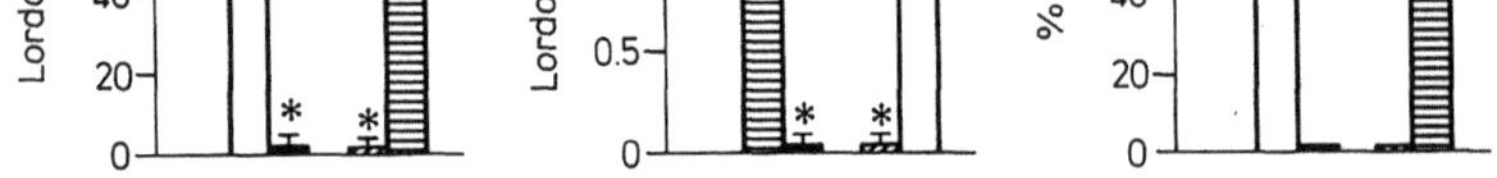

Fig. 6. The effects of hormonal manipulation during perinatal development on the activation of feminine reproductive behavior as measured by the lordosis quotient, lordosis quality, and proceptivity scores (% proceptive). Analysis of variance revealed a significant treatment effect of perinatal EB and ATD administration on the lordosis quotient ($F=573{,}7$; $P<0.001$) and the lordosis quality ($F=795.5$; $P<0.001$) scores. Comparisons of individual means using the Newman-Keuls test revealed that the lordosis quotient ($P<0.01$) and lordosis quality ($P<0.01$) scores of normal females and ATD-treated males differed significantly from those of normal males and EB-treated females. There were no significant differences between normal females and ATD-treated males, nor between normal males and EB-treated females

environment. Parsons and coworkers (1984) have found that organizational effects of testosterone via aromatization govern the apparently irreversible suppression of feminine reproductive behavior in the male, and lead to a supression of the capacity of the VMN of the hypothalamus to produce cytosolic progestin receptors in response to estrogen priming. The cytosolic progestin receptor content of the VMN in males was similar to that of females given perinatal estrogen, and significantly less than that of normal females and males given the aromatase inhibitor, ATD (Table 5). In addition, significantly less receptive and proceptive behavior was observed in males and females given perinatal EB than in normal females and in males given perinatal ATD (Fig. 6). These findings are consistent with the hypothesis that one event mediated by E_2 which underlies the activation of feminine reproductive behavior is the induction of cytosolic progestin receptors in the VMN. This capacity is apparently restricted by estrogen-mediated events in males during the perinatal period.

3.3 Temporal Correlations

The overall effects of P on a target tissue are determined partially by that target tissue's specific endocrine environment. Specifically, pretreatment with an estrogen appears to be the major requirement for optimal progestational responses in most tissues. In these target tissues, the role of estrogen is primarily that of inducing P sensitive target cells.

Oviduct and Uterus. P effects on biosynthesis in peripheral tissue are best understood in the chick oviduct and rodent uterus. At present, the oviduct is one of the few peripheral target tissues for which a protein, avidin, has been shown to be regulated specifically by P.

In the estrogen-withdrawn oviduct of the chick, the administration of P (2 mg) produced an increase in the number of initiated RNA chains within 30 min of administration (Table 6) (Chan et al. 1973). An increase in the activities of RNA polymerases I and II at 2 h (O'Malley et al. 1969) precedes the increase in $mRNA_{ovalbumin}$ and $mRNA_{avidin}$ molecules at 4 and 6 h, respectively (Chan et al.

Table 6. Biochemical effects of progesterone in oviduct

Parameter	Time first observed	References
Increase in RNA initiation sites	30 min	Chan et al. (1973)
Increase in the activities of RNA polymerase I and II	2 h	O'Malley et al. (1969)
Increase in $mRNA_{ovalbumin}$	4 h	Chan et al. (1973)
Increase in $mRNA_{avidin}$	6 h	Schwartz et al. (1979)
Increase in intracellular ovalbumin and avidin	6–72 h	Chan et al. (1973) O'Malley and Means (1974)

1973; Schwartz et al. 1979). Increases in intracellular avidin and ovalbumin are observed 6–72 h after P administration (Chan et al. 1973; O'Malley and Means 1974).

P effects on receptor regulation in peripheral tissue are best understood in the rodent uterus. Hsueh et al. (1976) demonstrated that P interferes with that phase of E_2 receptor replenishment which is dependent upon protein synthesis. This inhibition appears to be of physiological importance, as subsequent nuclear uptake of E_2 at 24 is depressed because of the decreased quantity of cytosolic receptor available for translocation (Hsueh et al. 1976). P also decreases the concentration of its own protein receptor in rodent uterus, within 24 h of administration (VuHai et al. 1977). P probably exerts its effect through an enhancement of the receptor inactivation rate, although the mechanism of this is not known (VuHai et al. 1977).

The minimal length of time which P must be retained by oviduct or uterine cells in order to promote the above changes in cellular metabolism has not been an area of intense investigation. O'Malley and coworkers (1969) observed that after subcutaneous administration of ^{3}H-P (sufficient to induce avidin synthesis), maximal uptake of tritium by the N-E-M (nuclear-myofibrillar-microsomal) fraction of the oviduct occurred between 30 and 60 min. Moreover, the oviduct rapidly metabolized this steroid: 1 h after P administration, only 35% of the radioactivity chromatographed as P (O'Malley et al. 1969). Based on these findings, one might reasonably conclude that retention of P by oviduct nuclei for a minimum of 6 h does not appear to be a prerequisite for this steroid to alter macromolecular synthesis at 24–28 h. However, additional data are necessary to define the period(s) of nuclear P receptor elevation which are sufficient to effect the subsequent changes in oviduct biosynthesis.

Brain and Pituitary. Pretreatment with estrogen also appears to be a major requirement for optimal progestational responses in the central nervous system. Very few events in the hypothalamus of the estrogen-primed rodent have been characterized with respect to their time of induction by P. In the E_2-treated rat, the administration of intravenous P can increase both the frequency and quality of the lordosis posture within 30–60 min (Lisk 1960; Meyerson 1972; Kubli-Garfias and Whalen 1977; McGinnis et al. 1981). In addition to facilitating sexual receptivity, P can activate components of sexual proceptivity (Beach 1942), such as hopping, darting, and ear-wiggling, within 30–60 min of intravenous administration (Lisk 1960; Meyerson 1972; Kubli-Garfias and Whalen 1977). If the P dose is sufficiently large, a refractory period follows, during which time animals are less sensitive to additional P (Zucker 1966; Blaustein and Wade 1977; Blaustein and Feder 1979; Fadem et al. 1979; Parsons and McEwen 1981; Parsons et al. 1981 b). This refractory period has been termed "sequential inhibition" by P (Powers and Moreines 1976).

The augmentation of positive feedback by P has been investigated with respect to time of induction. It should be noted that such augmentation may reflect P-induced changes in the hypothalamus and/or pituitary. When administered on late diestrus or proestrus, P increases LH serum levels within 6 h (Naller et al. 1956). In ovariectomized rodents treated with E_2, P can elevate plasma LH within

2–6 h of administration (Caligaris et al. 1971; DePaolo and Barraclough 1979; Wise et al. 1981). P also synergizes with E_2 to increase serum FSH levels in the ovariectomized rat. This elevation in serum FSH may be observed within several hours after P administration (Caligaris et al. 1971).

In contrast to P action in peripheral tissue, P does not modulate the concentration of the cytosolic estrogen receptor in the central nervous system (Marrone and Feder 1977). A decrease in cytosolic progestin receptor levels in the rat and guinea pig hypothalamus has been reported to occur 24 h after P administration (Blaustein and Feder 1979; Moguilewsky and Raynaud 1979; Schwartz et al. 1979; Parsons et al. 1981 b). Much of such a decrease may be attributed to a competition effect of the unlabeled P for ^{3}H-R5020 binding sites (Parsons et al. 1981 b). However, when one controls for competition by intracellular P, there is a 25%–35% decrease in hypothalamic, pituitary, and cortical cytosolic progestin receptor levels 24 h after P (2.5 mg) administration in the rat. In the guinea pig, an even larger decrease (55%) in hypothalamic cytosolic progestin receptors is observed 24 h after P (Blaustein 1982). These findings suggest that there is either a down-regulation of progestin receptors by P, or a delay in the recycling of the receptor from the nucleus to the cytoplasm (Parsons et al. 1981 b). Such a decrease may be of physiological importance, as subsequent nuclear uptake of P is depressed because of the decreased quantity of cytosolic progestin receptors available for translocation (Blaustein and Feder 1980; Blaustein 1982).

The minimal length of time which P must be retained by hypothalamic or pituitary cell nuclei in order to promote the above changes in feminine reproductive behavior or gonadotropin secretion has not been characterized. One recent study by McGinnis and colleagues has demonstrated that the rapid behavioral effects of intravenous P administration in E_2-primed rats are accompanied by rapid progestin receptor translocation in the MBH-POA. Furthermore, the facilitation of feminine reproductive behavior by P continues to be observed several hours after

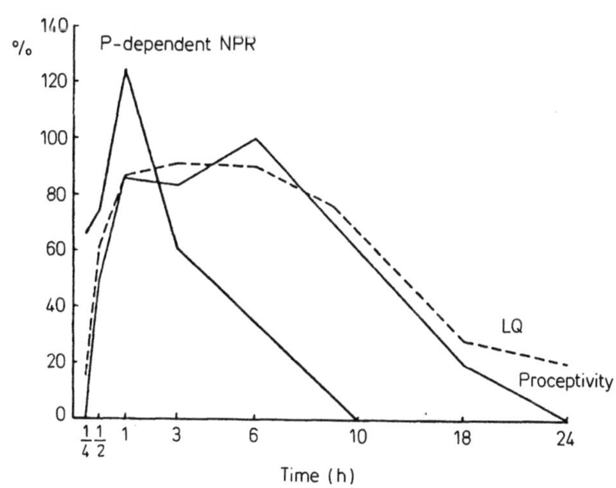

Fig. 7. Comparison between proceptivity and LQ scores, and nuclear ^{3}H-R5020 binding (*NPR*) expressed as percent elevation above control

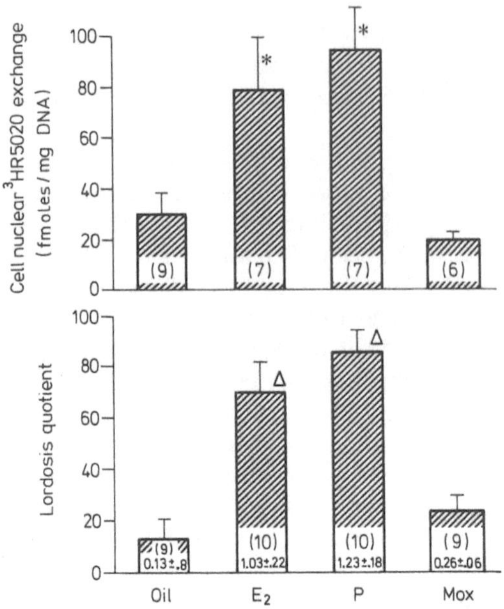

Fig. 8. Effect of acute treatment on hypothalamic nuclear progestin receptor (NPR) levels and sexual receptivity in ovariectomized rats primed with 100% E_2. Acute treatment consisted of 1 mg E_2, P, moxestrol (Mox), or oil vehicle. NPR levels in the MBH-POA are expressed as fmoles ^{3}H-R5020 bound/mg protein (*top panel*). Receptivity is presented as lordosis quotient (LQ) in bar graph and as lordosis quality score (LS) inside each bar. Data presented are means ± SEM. *Top panel:* One-way ANOVA significant, $F=6.53$; $P<0.005$.*, Duncan's test comparisons ($P<0.05$): E_2 and P > oil and Mox. *Bottom panel:* One-way ANOVA significant, $F=15{,}91$ (LQ) and $F=13{,}91$ (LS); $P<0.001$; △, Duncan's test comparisons ($P<0.01$): E_2 and P > oil and Mox (LQ and LS). Number of determinations (*exchange*) and number of rats (*LQ*) is indicated in *parentheses*

nuclear progestin receptors in the MBM-POA have returned to control levels (Fig. 7; McGinnis et al. 1981). These findings are consistent with the hypothesis that P-induced effects on feminine reproductive behavior involve genomic activation and subsequent protein synthesis.

In rats primed with low levels of E_2, a subsequent dose of E_2 (1 mg) can "mimic" P by elevating LQ and proceptivity scores within 3–5 h of subcutaneous administration (Kow and Pfaff 1975; Parsons et al. 1984). This increase in feminine reproductive behavior is preceded by an increase in brain and pituitary nuclear progestin receptor levels (Fig. 8; Parsons et al. 1984). A synthetic estrogen, moxestrol (Mox), which binds with considerably less affinity to the cytosolic progestin receptor, is unable to alter these parameters at an equivalent dose. In these respects, estrogen can elicit at least some of the acute actions of P in estrogen-primed animals, and probably does so by interacting with progestin receptors.

3.4 Functional Correlations

Oviduct. RNA and protein synthesis inhibitors have been employed to block P-induced events in the chick oviduct. O'Malley et al. (1969) found that the addition of actinomycin D to cultured oviduct cells pretreated with P produced a 90% inhibition of avidin synthesis, without inhibiting general protein synthesis. This observation suggests, but does not prove, that new DNA-dependent RNA synthesis is specifically required for the induction of avidin by P. Furthermore, O'Malley and coworkers found that the addition of cycloheximide to cultured oviduct cells pretreated with P also inhibited avidin synthesis, suggesting that the induction of avidin by P is a genomically mediated event requiring de novo protein synthesis.

Brain and Pituitary. The antibiotic anisomycin (ANI), has been shown to be effective in blocking protein synthesis in the central nervous system when administered systemically. Rainbow et al. (1981) found that ANI, when given 15 min prior to the administration of P, blocked the activation of proceptivity and the facilitation of receptivity in ovariectomized rats primed with estrogen. More recently, Rainbow and colleagues (1982 a) have demonstrated that stereotaxic application of ANI to the VMN also blocks the facilitation of reproductive behavior seen 4–6 h after subcutaneous P administration. Protein synthesis in the MBH was inhibited by 70% 4 h after intracranial ANI application, but no inhibition was observed in the POA corticomedial amygdala, or pituitary. Glaser and coworkers (1984) have demonstrated that the ANI blockade of estrous behavior of P is site-specific-VMN implants of ANI significantly decreased sexual behavior after P, whereas POA implants were without significant effect. These findings are consistent with, but not proof of, the hypothesis that P's activation of proceptivity and facilitation of receptivity in the female rat is a result of altered genomic expression and protein synthesis in the VMN.

In addition to facilitating sexual receptivity acutely, P may also play a role in the subsequent termination of reproductive behavior. As mentioned above, animals which receive P initially are less sensitive to additional P 24 h later than animals which did not receive initial P (sequential inhibition). A recent study by Par-

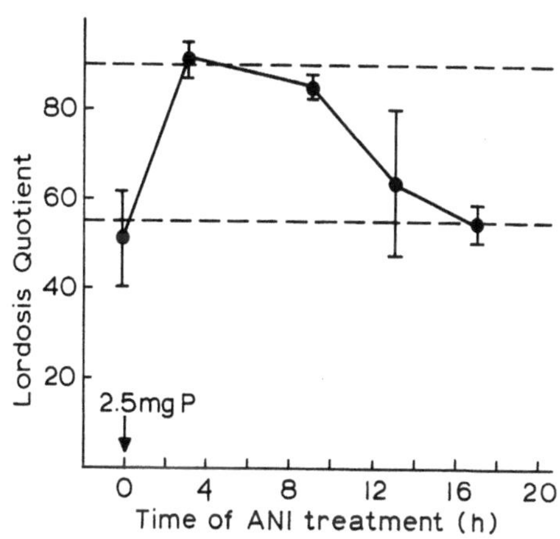

Fig. 9. Time course of effects of ANI on sequential inhibition. ANI was given as a single subcutaneous injection (100 mg/kg) at times indicated relative to an injection of 2.5 mg P in propylene glycol. Note that ANI at 3–10 h prevents sequential inhibition by P

sons and McEwen (1981) has demonstrated that administration of ANI, 15 min prior to and again 3 h after initial P treatment, can block the sequential inhibition of reproductive behavior. According to the time course of effective ANI application, the presumed protein synthesis which is responsible for the inhibition of sexual receptivity occurs between 3 and 13 h after the administration of P (Fig. 9; Parsons and McEwen 1981). The identification of the proteins associated with the sequential inhibition of receptivity, as well as those associated with the facilitation of reproductive behavior by P, remains an area for future investigation.

4 Interpretation and Outlook

Several central concepts have emerged from the study of estrogen- and P-mediated events in peripheral tissues and in the central nervous system. These concepts have addressed spatial, temporal, and functional aspects of either steroid-receptor interactions, or of steroid-induced events. A brief discussion of these ideas may serve to clarify the mechanism(s) by which estrogen and P synergize to induce feminine reproductive behavior and gonadotropin secretion.

Autoradiographic and physiological studies have identified descrete regions of estrogen and progestin receptor concentration in the vertebrate central nervous system. Estrogen receptors and estrogen-inducible progestin receptors are located primarily, but not exclusively, within a system of limbic, medial preoptic, and medial hypothalamic cells. These structures participate differentially in the regulation of a number of neuroendocrine end points, including the induction of gonadotropin secretion and feminine reproductive behavior. Experiments utilizing stereotaxic implantation of E_2 have demonstrated that the primary sites of E_2 action in the stimulation of LH release and the activation of the lordosis reflex are the medial POA and the VMN, respectively. Similar studies have indicated that the principal site of P action in the facilitation of receptivity and proceptivity is the VMN.

In peripheral tissue, it has been established that a number of steroid-induced events depend upon de novo protein synthesis. Several of these events, such as the induction of avidin synthesis by P in the chick oviduct, have been shown not only to be sensitive to blockade by protein synthesis inhibitors, but also to depend upon the synthesis of steroid-induced mRNAs. Similarly, the induction of gonadotropin secretion and feminine reproductive behavior have been shown to be sensitive to blockade by protein and RNA synthesis inhibitors, and the activation of feminine reproductive behavior has been shown to be correlated with the appearance of threshold levels of inducible progestin receptors in the MBH-POA. Thus, at least some of the permissive events in the induction of gonadotropin secretion and the activation of feminine reproductive behavior by steroid hormones require genomic participation.

Steroid-induced events which depend on de novo protein synthesis require that the steroid-receptor complex be present in the cell nucleus, in order for transcription to be modified. However, presence of steroid hormone-receptor complexes in appropriate target cell nuclei is insufficient to stimulate events such as true uterine growth, LH release, and the activation of the lordosis reflex. In order for these physiological responses to be elicited, estrogen-receptor complexes must be retained by target cell nuclei for specified time periods. Continuous occupation of nuclear estrogen receptors may not be necessary for the stimulation of late uterotrophic responses, LH release, and the activation of feminine reproductive behavior; experimental evidence suggests that nuclear estrogen receptors must be elevated at both an early and a late period in order to effect these events.

A tentative conclusion which emerges from a number of studies is that estrogen may initiate at least two distinct phases of protein synthesis in uterine, pituitary, and hypothalamic cells, which culminate in uterine cell division, the release of LH from pituitary, and the activation of lordosis. These alterations in neuroendocrine function occur 18–30 h after the initiation of estrogen treatment. Consistent with such relatively long latent periods is the "stability" of at least one protein associated with the induction of receptivity by E_2. The half-life of the estrogen-inducible progestin receptor in the MBH-POA and pituitary is approximately 24 h; such stability may be characteristic of proteins associated with the activation of behavior by E_2.

P stimulates several known biosynthetic events in peripheral tissue, facilitates gonadotropin secretion and the lordosis reflex, and activates proceptive behaviors in the rodent. Retention of P for a number of hours does not appear to be a prerequisite for P to stimulate avidin synthesis in oviduct nuclei, nor for it to activate proceptivity in the female rat. In fact, when P is administered intravenously, receptive and proceptive behaviors in the female rat are facilitated within 30 min. Experiments employing the protein synthesis inhibitor, ANI, suggest that at least some of the essential proteins associated with P's facilitation of behavior turn over more rapidly than those associated with the activation of mating by E_2.

Our understanding of the controls on feminine reproductive behavior and gonadotropin secretion by E_2 and P has been facilitated by the exploration of spatial, temporal, and functional correlates of steroid hormone receptors in the vertebrate brain. It is hoped that investigators will continue to elucidate those biochemical and electrical changes in the central nervous system which are related causally to alterations in neuroendocrine function.

References

Anderson JN, Clark JH, Peck EJ (1972) Oestrogen and nuclear binding sites. Biochem J 126:561–567

Arai Y, Gorski RA (1968) Effectofante-estrogen on steroid-inducal sexual receptivity in ovariectomized rats. Physiol. Behav. 3:351–353

Arnold A, Nottebohm F, Pfaff DW (1976) Hormone-concentrating cells in vocal control and other areas of the brain of the zebra finch (poephila quitata). J Comp Neurol 165:487–512

Ball J (1937) A test for measuring sexual excitability in the female rat. Comp Psychol Mon 14:1–37

Balthazart J, Blaustein JD, Cheng MT, Feder HH (1980) Hormones modulate the concentration of cytoplasmic progestin receptors in the brain of male ring dove (streptopelia risoria). J Endocrinol 86:251–261

Bapra J, Neff NH, Costaj E (1971) A method for studying norepinephrine and serotonin metabolism in small regions of the rat brain: effect of ovariectomy on amine metabolism in anterior and posterior hypothalamus. Endocrinology 89:1345–1349

Barfield RJ, Cheng JJ (1977) Activation of estrous behavior in ovariectomized rats by intracerebral implants of estradiol benzoate. Endocrinology 101:1716–1725

Barfield RJ, Ronay G, Pfaff DW (1978) Autoradiographic localization of androgen concentrating cells in the brain of the male domestic fowl. Neuroendocrinology 26:297–311

Barker KL, Warren JD (1966) Template capacity of uterine chromatin: control by estradiol. Proc Natl Acad Sci USA 56:1298–1302

Barnea A, Gorski J (1970) Estrogen-induced protein. Time course of synthesis. Biochemistry 9:1899–1904

Beach FA (1942) Importance of progesterone to induction of sexual receptivity in spayed female rats. Proc Soc Exp Biol Med 51:369–371

Biegon A, McEwen BS (1982) Modulation by estradiol of serotonin receptors in brain. J Neurosci 2:199–205

Blaustein JD (1982) Alteration of sensitivity to progesterone facilitation of lordosis in guinea pigs by modulation of hypothalamic progestin receptors. Brain Res 243:287–300

Blaustein JD, Feder HH (1979) Cytoplasmic progestin receptors in the female guinea pig brain and their relationship to refractoriness in expression of female sexual behavior. Brain Res 177:489–498

Blaustein JD, Feder HH (1980) Nuclear progestin receptors in guinea pig brain measured by an in vitro exchange assay after hormonal treatments that affect lordosis. Endocrinology 106:1061–1069

Blaustein JD, Wade GW (1977) Sequential inhibition of sexual behavior by progesterone in female rats: comparison with a synthetic antiestrogen. J Comp Physiol Psychol 91:752–760

Blaustein JD, Dudley SD, Grey JM, Roy EJ, Wade GW (1979) Long-term retention of estradiol by brain cell nuclei and female rat sexual behavior. Brain Res 173:355–359

Bueno J, Pfaff DW (1976) Single unit recording in hypothalamus and preoptic area of estrogen-treated and untreated ovariectomized female rats. Brain Res 101:67–68

Caligaris L, Astrada JJ, Taleisnik S (1971) Biphasic effect of progesterone on the release of gonadotropin in rats. Endocrinology 89:331–337

Cardinali DP, Gomez E (1977) Changes in hypothalamic noradrenaline, dopamine and serotonin uptake after estradiol administration to rats. J Endocrinol 73:181–182

Chan L, Means AR, O'Malley BW (1973) Rates of induction of specific translatable messenger RNAs for ovalbumin and avidin by steroid hormones. Proc Natl Acad Sci USA 70:1870–1874

Clark JH, Peck EJ (1976) Nuclear retention of receptor-estrogen complex and nuclear receptor sites. Nature 260:635–636

Clemens LG, Shryne J, Gorski RA (1970) Androgen and development of progesterone and responsiveness in male and female rats. Physiol Behav 5:673–679

Davidson JM, Levine S (1969) Progesterone and heterotypical sexual behavior in male rats. J Endocrinol 44:129–137

Davis PG, McEwen BS, Pfaff DW (1979) Localized behavioral effects of tritiated estradiol implants in the ventromedial hypothalamus of female rats. Endocrinology 104:893–903

DeAngelo AB, Gorski J (1970) Role of RNA synthesis in the estrogen induction of a specific uterine protein. Proc Natl Acad Sci USA 66:693–700

Debeljuk L, Rettori V, Rozados RV, Velez CV (1975) Effect of anisomycin D and estradiol on the response to LH-releasing hormone in neonatally androgenized female rats. Proc Soc Exp Biol Med 150:229–231

DeLean A, Goron M, Kelly PA, Labrie F (1977) Changes of pituitary thyrotropin releasing hormone (TRH) receptor level and prolactin response to TRH during the rat estrous cycle. Endocrinology 100:1505–1510

DePaolo L, Barraclough CA (1979) Dose dependent effects of progesterone on the facilitation and inhibition of spontaneous gonadotropin surges in estrogen-treated ovariectomized rats. Biol Reprod 21:1015–1023
Drouin J, LaGace L, Labrie F (1976) Estradiol-induced increase of the LH releasing hormone (LHRH) in rat anterior pituitary cells in culture. Endocrinology 99:1477–1481
Eilon G, Gorski J (1972) Nuclear estrogen receptors: accumulation and retention as influenced by exogenous and endogenous estrogens. Fed Proc 31:245
Fadem BH, Barfield RJ, Whalen RE (1979) Dose-response and time-response relationships between progesterone and the display of patterns of receptive and proceptive behavior in the female rat. Horn Behav 13:40–48
Feder HH, Marrone BL (1977) Progesterone: its role in the central nervous system as a facilitator and inhibitor of sexual behavior and gonadotropin release. Ann NY Acad Sci 286:331–333
Floody O, Pfaff DW (1974) Steroid hormones and aggressive behavior: approaches to the study of hormone-sensitive mechanisms for behavior. Res Publ Assoc Res Nerv Ment Dis 52:149–185
Gerlach JL, McEwen BS, Pfaff DW, Moskovitz S, Ferin M, Carmel PW, Zimmerman EA (1975) Cells in regions of rhesus monkey and pituitary retain radioactive estradiol, corticosterone and cortisol differentially. Brain Res 103:603–612
Glaser JH, Cislo SC, Barfield RJ (1984) Intracerebral anisomycin blockade of progesterone facilitated estrous behavior in rats is site specific. Neuroendocrinology (to be published)
Glasser SR, Chytil F, Spelsberg TC (1972) Early effects of estradiol-17-β on the chromatin and activity of deoxyribonucleic acid-dependent ribonuclease acid polymerases in the rat uterus. J Biochem 130:947–957
Goodman RL (1978) The site of positive feedback action of estradiol in the rat. Endocrinology 102:151–159
Gorski J (1964) Early estrogen effects on the activity of uterine ribonucleic acid polymerase. J Biol Chem 239:889–892
Goy RW, McEwen BS (1980) Sexual differentiation of the brain. MIT, Cambridge, pp 211
Green R, Luttge WG, Whalen RE (1970) Induction of receptivity in ovariectomized female rats by a single intravenous injection of estradiol-17 β. Physiol Behav 5:137–141
Hamilton TH (1963) Isotopic studies on estrogen-induced accelerations of ribonucleic acid and protein synthesis. Proc Natl Acad Sci USA 49:373–379
Hardy DF (1972) Sexual behavior in continuously cycling rats. Behavior 41:288–297
Hardy D (1976) Sexual attractivity, proceptivity and receptivity in female mammals. Horm Behav 7:105–138
Hardy DF, DeBold JF (1977) The relationship between levels of exogenous hormones and the display of lordosis by the female rat. Horm. Behav. 9:380–386
Harris J, Gorski J (1978) Discontinous requirement for estrogen in stimulation of DNA synthesis in immature rat uterus. Endocrinology 103:240–246
Heil H, Meltzer V, Kuhil H, Abraham R, Taubert HD (1971) Stimulation of L-cystine-aminopeptidase activity by hormonal steroids and steroid-analogs in the hypothalamus and other tissues of the female rat. Fertil Steril 22:181–187
Hisaw FL (1959) Comparative effectiveness of estrogens on fluid inhibition and growth of the rat's uterus. Endocrinology 64:276–289
Hsueh AJW, Peck EJ, Clark JH (1976) Control of uterine estrogen receptor levels by progesterone. Endocrinology 98:438–444
Jackson GL (1972) Effect of actinomycin D on estrogen-induced release of luteinizing hormone in ovariectomized rats. Endocrinology 91:1284–1287
Jackson GL (1973) Time interval between injection of estradiol benzoate and LH release in the rat and effect of actinomycin D or cycloheximide. Endocrinology 93:887–891
Jacobi J, Lloyd HM, Mearnes JD (1977) Onset of estrogen-induced prolactin secretion and DNA synthesis by the rat pituitary gland. J Endocrinol 72:35–39
Jensen et al.

Johnson PG, Davidson JM (1979) Priming action of estrogen: minimum duration of exposure for feedback and behavioral effects. Neuroendocrinology 28:155–159

Kalmer SM, Lowenstein JM (1958) The effect of estrogens on the uptake of albumin and electrolytes by the rat uterus. J Pharmacol Exp Ther 122:163–167

Kalra S (1975) Observations on facilitation of the preovulatory rise of LH by estrogen. Endocrinology 96:23–28

Kalra SP (1976) Tissue levels of luteinizing hormone-releasing hormone in the preoptic area and hypothalamus, and serum concentrations of gonadotropins following anterior hypothalamic deafferentation and estrogen treatment of the female rat. Endocrinology 99:101–107

Kato J, Grouch T (1977) Specific progesterone receptors in the hypothalamus and anterior hypophysis of the rat. Endocrinology 101:912–928

Katzenellenbogen BS, Gorski J (1975) Estrogen action on syntheses of macromolecules in target cells. In: Litwack G (ed) Biochemical actions of hormones, vol 3. Academic, New York, pp 187–243

Kaye AM, Sheratzky D, Lindner HR (1972) Kinetics of DNA synthesis in immature rat uterus: age, dependence and estradiol stimulation. Biochim Biophys Acta 261:475–486

Kelley DB, Pfaff DW (1975) Locations of steroid hormone-concentrating cells in the central nervous system of *Rana pipiens*. Soc Neurosci Abstr 438

Kelley DB, Pfaff DW (1978) Generalizations from comparative studies on neuroanatomical and endocrine mechanisms of sexual behavior. In: Hutchison J (ed) Biological determinants of sexual behavior. Wiley, Chicester, pp 225–254

Kelley DB, Morell JI, Pfaff DW (1975) Autoradiographic localization of hormone-concentrating cells in the brain of an amphibian, *Xenopus laevis*. I. Testosterone. J Comp Neurol 164:47–62

Kelner KL, Miller AL, Peck EJ (1980) Estrogens and the hypothalamus: nuclear receptor and RNA polymerase activation. J Recept Res 1:215–227

King WJ, Greene GL (1984) Monoclonal antibodies localize oestrogen receptor in the nuclei of target cells. Nature 307:745–747

Kow LM, Pfaff DW (1975) Induction of lordosis in female rats: two modes of estrogen action and the effect of adrenalectomy. Horm Behav 6:259–276

Krey LC, Parsons B (1982) Characterization of estrogen stimuli sufficient to initiate cyclic luteinizing hormone release in acutely ovariectomized rats. Neuroendocrinology 34:315–322

Kubli-Garfias C, Whalen RE (1977) Induction of lordosis behavior in female rats by intravenous administration of progestins. Horm Behav 9:380–386

Kuehl FA, Ham EA, Zanetti ME, Sanford C, Nicol SE, Goldberg ND (1974) Estrogen-related increases in uterine guanosine 3′:5′ cyclic monophosphate levels. Proc Natl Acad Sci USA 71:1866–1870

Kuehn RF, Beach FA (1963) Quantitative measurement of sexual receptivity in female rats. Behavior 21:282–299

Lan NC, Katzenellenbogen BS (1976) Temporal relationships between hormone receptor binding and biological responses in the uterus: studies with short- and long-acting derivatives of estradiol. Endocrinology 98:220–227

Lisk RD (1960) A comparison of the effectiveness of intravenous, as opposed to subcutaneous, injections of progesterone for the induction of estrous behavior in the rat. Can J Biochem Physiol 38:1381–1383

Litteria M, Thorner MW (1974) Alterations in the incorporation of ^{3}H-lysine into proteins of the medial preoptic area and specific hypothalamic nuclei after ovariectomy in the adult female rat. J Endocrinol 60:377–378

Liu TC, Jackson GL (1977) Effect of in vivo treatment of estrogen on luteinizing hormone synthesis and release by rat pituitaries in vitro. Endocrinology 100:1294–1302

Luine VN, McEwen BS (1976) Steroid hormone receptors in brain and pituitary: topography and possible functions. In: Goy RW, Pfaff DW (eds) Sexual Behavior. Plenum, New York

Luine VN, McEwen BS (1977) Effects of an estrogen antagonist on enzyme activities and ^{3}H-estradiol nuclear binding in uterus, pituitary and brain. Endocrinology 100:903–910

Luine VN, Khylehevskaya RI, McEwen BS (1974) Estrogen effects on brain and pituitary enzyme activities. J Neurochem 23:925–934

Luine VN, Khylchevskaya RI, McEwen BS (1975) Effect of gonadal hormones on enzyme activities in brain and pituitaries of male and female rats. Brain Res 86:283–292

MacLusky NJ, McEwen BS (1978) Oestrogen modulates progestin receptor concentration in some brain regions but not in others. Nature 274:276–277

MacLusky NJ, McEwen BS (1980a) Progestin receptors in the developing rat brain and pituitary. Brain Res. 189:262–268

MacLusky NJ, McEwen BS (1980b) Progestin receptors in rat brain: distribution and properties of cytoplasmic progestin binding sites. Endocrinology 106:192–202

Malsbury C, Kow L-M, Pfaff DW (1977) Effects of sexual receptivity in the hormone-primed female hamster by electrical stimulation of the medial preoptic area. Physiol Behav 19:223–237

Markaverich B, Clark J, Hardin (1978) RNA transcriptum and uterine growth. *Biochemistry* 17:3146–3152

Marrone BL, Feder HH (1977) Characteristics of ^{3}H-estrogen and ^{3}H-progestin uptake and effects of progesterone on ^{3}H-estrogen uptake in brain, anterior pituitary, and peripheral tissues of male and female guinea pigs. Biol Reprod 17:42–57

Martin L, Motia M, Fraschin F (1970) The hypothalamus. Academic, New York, p 705

Mastro A, Hymer WC (1973) The effects of age and estrone treatment on DNA polymerase activity in anterior pituitary glands of male rats. J Endocrinol 59:107–109

McEwen BS, Gerlach J, Micco DJ (1975a) Putative glucocorticoid receptors in hippocampus and other regions of the rat brain. In: Isaacson RL, Pribram KH (eds) The hippocampus, structure and development, vol 1. Plenum, New York, pp 285–322

McEwen BS, Pfaff DW, Chaptal C, Luine VN (1975b) Brain cell nuclear retention of ^{3}H-estradiol doses able to promote lordosis: temporal and regional aspects. Brain Res 86:155–161

McEwen BS, Davis PG, Parsons B, Pfaff DW (1979) The brain as a target for steroid hormone action. Annu Rev Neurosci 2:65–112

McGinnis MY, Parsons B, Rainbow TC, Krey LC, McEwen BS (1981) Temporal relationship between cell nuclear progestin receptor levels and sexual receptivity following intravenous progesterone administration. Brain Res 218:365–371

Meinkoth J, Quadagno D, Bast JP (1979) Depression of steroid-induced sex behavior in the ovariectomized rat by intracranial injection of cycloheximide: preoptic area compared to the ventromedial hypothalamus. Horm Behav 12:199–204

Mester J, Baulieu EE (1975) Dynamics of oestrogen-receptor distribution between the cytosol and nuclear fractions of immature rat uterus after oestradiol administration. 146:617–623

Meyerson B (1972) Latency between intravenous injection of progestins and the appearance of estrous behavior in estrogen-treated ovariectomized rats. Horm Behav 3:1–9

Miller WL, Knight MM, Grimek HJ, Gorski J (1977) Estrogen regulation of follicle stimulating hormone in cell cultures of sheep pituitaries. Endocrinology 100:1306–1316

Moguilewsky M, Raynaud J-P (1979) The relevance of hypothalamic and hypophyseal progestin receptor regulation in the induction and inhibition of sexual behavior in the female rat. Endocrinology 105:516–522

Morin LP, Feder HH (1974) Inhibition of lordosis behavior in ovariectomized guinea pigs by mesencephalic implants of progesterone. Brain Res 70:71–80

Morrell JI, Pfaff DW (1975) Autoradiographic localization of ^{3}H-estradiol uptake in the brain of the female mink, *mustela vision*. Anat Rev 181:430

Morrell J, Pfaff DW (1981) Autoradiographic technique for steroid hormone localization: application to the vertebrate brain. In: Adler N (ed) Neuroendocrinology of reproduction: physiology and behavior. Plenum, New York, pp 519–531

Morrell J, Kelley D, Pfaff DW (1975) Sex steroid binding in the brains of vertebrates: studies with light microscopic autoradiography. In: Krigge K, Scott D, Kobayashi H, Ishii S (eds) Brain-endocrine interaction, vol 2. Karger, Basel, pp 230–256

Moyer KE (1976) The psychobiology of aggression. Harper and Row, New York, p 402

Notides AC, Gorski J (1966) Estrogen-induced synthesis of a specific uterine protein. Proc Natl Acad Sci USA 56:230–235

O'Malley B (1967) In vitro hormonal induction of a specific protein (avidin) in chick oviduct. Biochemistry 6:2546–2551

O'Malley BW, Means AR (1974) Female steroid hormones and target cell nuclei. Science 183:610–620

O'Malley BW, McGuire WL, Kohler PO, Korenman SG (1969) Studies on the mechanism of steroid hormone regulation of synthesis of specific proteins. Recent Prog Horm Res 25:105–152

Parsons B, McEwen BS (1981) Sequential inhibition of sexual receptivity by progesterone is prevented by a protein synthesis inhibitor and is not causally related to decreased levels of hypothalamic progestin receptors in the female rat. J Neurosci 1:527–531

Parsons B, MacLusky NJ, Krey LC, Pfaff DW, McEwen BS (1980) The temporal relationship between estrogen-inducible progestin receptors in the female rat brain and the time course of estrogen activation of mating behavior. Endocrinology 107:774–779

Parsons B, Rainbow TC, Pfaff DW, McEwen BS (1981 a) Oestradiol, sexual receptivity and cytosol progestin receptors in the hypothalamus. Nature 292:58–59

Parsons B, McGinnis MY, McEwen BS (1981 b) Sequential inhibition by progesterone: effects on sexual receptivity and associated changes in brain cytosol progestin binding in the female rat. Brain Res 221:149–160

Parsons B, McEwen BS, Pfaff DW (1982 a) A discontinuous schedule of estradiol treatment is sufficient to activate progesterone-facilitated feminine reproductive behavior and to increase cytosol progestin receptor levels in the hypothalamus of the rat. Endocrinology 110:613–619

Parsons B, Rainbow TC, Pfaff DW, McEwen BS (1982 b) Hypothalamic protein synthesis essential for the activation of the lordosis reflex in the female rat. Endocrinology 110:620–624

Parsons B, Rainbow TC, MacLusky NJ, McEwen BS (1982 c) Progestin receptor levels in rat hypothalamic and limbic nuclei. J Neurosci 2:1446–1452

Pfaff DW (1968) Uptake of estradiol-17-H^3 in the female rat brain. An autoradiographic study. Endocrinology 82:1145–1149

Pfaff DW, Keiner M (1973) Atlas of estradiol-concentrating cells in the central nervous system of the female rat. J Comp Neurol 151:121–158

Pfaff DW, Moskovitz S, Gerlach J, McEwen B, Carmel P, Ferin M, Zimmerman E (1974) Autoradiographic localization of cells which bind estradiol or corticosterone in the brain and pituitary of the female rhesus monkey. Proc. Int Cong Physiol Sci 26:281

Pietras RJ, Szego CM (1976) Endometrial cell calcium and oestrogen action. Nature 253:357–359

Powers JB, Moreines J (1976) Progesterone: examination of its postulated inhibitory actions on lordosis during the rat estrus cycle. Physiol Behav 17:492–498

Puig-Duran E, MacKinnon PCB (1978) Ontogeny of prolactin and luteinizing hormone responses to estrogen and progesterone in immature rats. J Endocrinol 76:321–331

Quadagno DM, Ho GK (1975) The reversible inhibition of steroid-induced sexual behavior by intracranial cycloheximide. Horm Behav 6:19–26

Rainbow TC, DeGroff V, Luine VN, McEwen BS (1980) Estradiol-17-β increases the number of muscarinic receptors in hypothalamic nuclei. Brain Res 198:239–243

Rainbow TC, Davis PG, McEwen BS (1981) Anisomycin inhibits the activation of sexual behavior by estradiol and progesterone. Brain Res 194:548–555

Rainbow TC, McGinnis MY, Davis PG, McEwen BS (1982 a) Application of anisomycin to the lateral ventromedial nucleus of the hypothalamus inhibits the activation of sexual behavior by estradiol and progesterone. Brain Res 233:417–423

Rainbow TC, Parsons B, McLusky NJ, McEwen BS (1982 b) Estradiol receptor levels in rat hypothalamic and limbic nuclei. J Neurosci 2:1439–1445

Robinson JA, Leavitt WW (1971) Estrogen-related changes in anterior pituitary RNA levels. Proc Soc Exp Biol Med 139:471–475

Rubin BS, Barfield RJ (1980) Priming of estrous responsiveness by implants of 17-β-estradiol in the ventromedial hypothalamic nucleus of female rats. Endocrinology 106:504–509

Rubin BS, Barfield RJ (1984) Progesterone in the ventromedial hypothalamus facilitates estrous behavior in ovariectomized, estrogen-primed rats. Endocrinology 113:797–805

Sar M, Stumpf WE (1973) Neurons of the hypothalamus concentrate ^{3}H-progesterone or its metabolites. Science 182:1266–1268

Sarff M, Gorski J (1971) Control of estrogen-binding protein concentration under basal conditions and after estrogen administration. Biochemistry 10:2557–2563

Schwartz SM, Blaustein JD, Wade GW (1979) Inhibition of estrous behavior by progesterone in rats: role of neural estrogen and progestin receptors. Endocrinology 105:1078–1082

Shelesnyak MC (1959) Effect of estrogens on histamine release in the rat uterus. Proc Soc Exp Biol Med 100:739–742

Smith DE, Gorski J (1968) Estrogen control of uterine glucose metabolism: an analysis based on the transport and phosphorylation of 2-deoxyglucose. J Biol Chem 243:4169–4174

Smith ER, Barker KL (1974) Effects of estradiol and nicotinamide adenine dinucleotide phosphate on the rate of synthesis of uterine glucose 6-phosphate dehydrogenase. J Biol Chem 249:6541–6547

Smith MS, Freeman ME, Neill JD (1975) The control of progestin secretion during the estrous cycle and early pseudo-pregnancy in the rat: prolactin, gonadotropin and steroid levels associated with resue of the corpus luteum of pseudopregnancy. Endocrinology 96:219–226

Sodersten P (1975) Receptive behavior in developing female rats. Horm Behav 6:307–317

Sodersten P, Eneroth P, Hansen S (1981) Induction of sexual receptivity in ovariectomized rats by pulse administration of oestradiol-17-β. J Endocrinol 89:55–62

Stumpf WE (1968) Estradiol-concentrating neurons: topography in the hypothalamus by dry mount autoradiography. Science 162:1001–1003

Szego CM, Davis JS (1967) Adenosine 3′5′ monophosphate in rat uterus; acute elevation by estrogen. Proc Natl Acad Sci USA 58:1711–1718

Szego CM, Roberts S (1953) Steroid action and interaction in uterine metabolism. Recent Prog Horm Res 8:419–469

Tang LKL, Spies HG (1975) Effects of gonadal steroids on the basal and LRF-induced gonadotropin secretion by cultures of rat pituitary. Endocrinology 96:349–386

Tennet BJ, Smith ER, Davidson JM (1980) The effects of estrogen and progesterone on female rat proceptive behavior. Horm Behav 14:65–75

Terkel AS, Shryne J, Gorski RA (1973) Inhibition of estrogen facilitation of sexual behavior by the intracerebral influence of actinomycin D. Horm Behav 4:377–386

Tobarz RR, Crews DP, McEwen BS (1981) Estrogen-sensitive progestin binding sites in the brain of the lizard, *anolis carolinensis*. Brain Res 220:95–105

Vilchez-Martinez JA, Animura A, Debeljuk L, Schally AV (1974) Biphasic effect of estradiol benzoate on the pituitary responsiveness to LH-RH. Endocrinology 94:1300–1303

VuHai MT, Logeat F, Warembourg M, Milgrom E (1977) Hormonal control of progesterone receptors. Ann NY Acad Sci 286:199–209

Ward LL (1972) Prenatal stress feminizes and demasculinizes the behavior of males. Science 175:82–84

Warembourg M (1975a) Radioautographic study of the rat brain after injection of (1,2-M^3) corticosterone. Brain Res 89:61–70

Warembourg M (1975b) Radioautographic study of the rat brain and pituitary after injection of ^{3}H-dexamethasone. Cell Tiss Res 161:183–191

Warembourg M (1978) Uptake of ^{3}H-labeled synthetic progestin by rat brain and pituitary. A radiographic study. Neurosci Lett 7:1–8

Welshons WV, Lieberman ME, Gorski J (1984) Nuclear localization of unoccupied oestrogen receptors. Nature 307:747–749

Whelly SM, Barker KL (1974) Early effect of estradiol on the peptide elongation rate by uterine polyribosomes. Biochemistry 12:341–346

Williams CL (1979) Steroids induce lordosis and ear wiggling in infant rats. Abstracts, Eastern Conference on Reproductive Behavior, New Orleans

Wise PM, Rance N, Barraclough CA (1981) Effects of estradiol and progesterone on catecholamine turnover rates in discrete regions of hypothalamus in ovariectomized rats. Endocrinology 108:2186–2193

Ydstebo R, Sodersten P (1977) Induction of sexual receptivity in ovariectomized female rats by subcutaneous implants of estradiol-17-β. Horm Behav 9:130–140

Zigmond RE, Nottebohm F, Pfaff DW (1973) Androgen-concentrating cells in the midbrain of a songbird. Science 179:1005–1007

Zucker I (1966) Facilitatory and inhibitory effects of progesterone on sexual responses of spayed guinea pigs. J Comp Physiol Psychol 62:376–381

Estradiol-Progesterone Interactions in the Reproductive Behavior of Female Rats

P. Södersten [1]

Contents

1 Introduction

Experiments initiated by W. C. Young and collaborators in the early 1930s (reviewed by Young 1961, 1969) have dominated the work on control of feminine mating behavior by ovarian hormones. The conceptual developments of this group still guide research in the field as well as in related fields, e.g., the hormonal control of body weight regulation (Wade 1976). In initial attempts to identify the endocrine factors controlling the induction of sexual receptivity in the guinea pig, it was found that growth of the Graafian follicle was accelerated in the hours

1 Department of Psychiatry, Karolinska Institute, S-14186 Huddinge, Sweden

Current Topics in Neuroendocrinology, Vol 5

preceding the onset of sexual behavior (Myers et al. 1936). Knowing that exogenous administration of luteinizing hormone (LH) induces rapid growth of the preovulatory follicle (Foster and Hisaw 1935), Dempsey et al. (1936) found that administration of LH to cyclic guinea pigs induced sexual receptivity if given at a stage of the estrous cycle when the ovaries contained ripe follicles; but not if given early in the cycle or if given to ovariectomized animals pretreated with estrogen. It was shown that the LH acted on the well-developed ovarian follicles, stimulating the production of a factor which in turn induced the behavior. The follicular factor was identified as progesterone. Thus, the hormonal factors controlling the display of sexual behavior during the guinea pig estrous cycle were shown to be estrogen and progesterone, and it was shown that progesterone acted on a background of estrogen priming, i.e., after a period of estrogen conditioning. A second injection of progesterone, in ovariectomized animals previously brought into sexual receptivity by estrogen and progesterone, was not followed by another period of receptivity (Boling et al. 1938); however, ovariectomized animals could be repeatedly made receptive during a 15-day period by sequential administrations of estrogen and progesterone. In the intact animal, on the other hand, repeated behavioral heats could not be elicited by the estrogen-progesterone treatment. Thus, the animals did not readily display sexual behavior in response to the estrogen-progesterone treatment if this was administered during the luteal phase, i.e., prior to day 12 of the cycle. However, most animals responded to hormone treatment given after day 12. Accordingly, it was suggested that the presence of a functional corpus luteum might counteract the conditioning action of estrogen (Dempsey et al. 1936; Goy et al. 1966).

These experiments have provided the basis for almost all of the subsequent work on the roles of estrogen and progesterone in the induction and termination of feminine sexual behavior. Both facilitatory and inhibitory effects of progesterone on the behavior of estrogen-treated animals were postulated, and it was suggested that the inhibitory action of progesterone may be important for termination of the behavior (Goy et al. 1966). Reasonably recent reviews of these behavioral effects of progesterone, using the guinea pig as the model, have appeared (Feder and Marrone 1977; Morin 1977). The emphasis here will be on recent experiments on the rat.

2 The Sexual Behavior of the Female Rat

The female rat has a brief estrous cycle, of 4 or 5 days duration, which lacks a functional luteal phase. Thus, every 4th or 5th day the rat comes into behavioral estrus and ovulates. The onset of sexual behavior occurs at the time of lights-off in a colony room with a standard 12:12 h light:darkness (LD) cycle, and all rats are receptive some 4 h after lights-off (Södersten and Eneroth 1981 a). The rats remain sexually receptive for about 13 h and then gradually cease exhibiting receptive behavior (Södersten and Eneroth 1981 a). Ever since it was introduced (Beach 1942), the lordosis quotient (LQ: the number of lordosis responses in response to the mounting male, i.e., concave back flexion, lateral tail deviation, and neck extension; divided by the number of mounts by the stimulus male) has remained the most commonly used measure of female rat sexual behavior. The LQ yields reliable quantitative data on the consummatory aspect of female behavior but does not take the motivational aspects into consideration. An alternative method commonly used with the guinea pig is the manual stimulation technique (Zucker 1967a), in which the experimenter manually stimulates the animal with the middle and index fingers bilaterally on the flanks, followed by combined flank perineum palpation. This stimulation mimics the stimulation normally provided by the male rat during copulation (Pfaff et al. 1977). Like the LQ test, the manual stimulation technique ignores the motivational aspects of female rat sexuality and, also, it yields only results of an all-or-none character. Yet the procedure may be useful, depending upon the aim of the study: it provides results which correlate well with those from LQ tests (Södersten and Hansen 1977), and the copulation-induced inhibition of lordosis (Lodder and Zeilmaker 1976) is avoided (Zucker 1967a). Under certain experimental conditions, however, the procedure may be less useful (Adler and Bell 1969).

In comparison to the amount of work carried out in rats on the consummatory aspect of female sexual behavior, i.e., the lordosis response and its hormonal control mechanisms, considerably less attention has been paid to the development of measures of female sexual motivation; although detailed accounts of the behavior of sexually receptive rats are available (McClintock and Adler 1978). This was emphatically pointed out by Beach (1976), who introduced the term proceptivity to cover these aspects of female rat sexuality. However, this may not be surprising since the display of proceptive behaviors by sexually receptive female rats, i.e., ear-wiggling, hopping, and darting movements, seems to be controlled by the same hormonal factors as lordosis. Thus, lordosis can be induced in ovariectomized rats by estrogen in the absence of progesterone, and increments of the estrogen dose can stimulate proceptivity (Zemlan and Adler 1977). When estrogen is administered in a dose which results in a low level of lordosis responding and progesterone is added, an increase in the LQ ist observed and, correlatively, the display of proceptive behaviors increases (Beach 1942; Whalen 1974; Tennent et al. 1980; Edwards and Pfeifle 1983). Also, behaviors indicative of a low sexual motivation, such as kicking off the male who attempts to mount, which may be seen in estrogen-primed animals, are reduced by the administation of progesterone (Beach 1942). Progesterone obviously stimulates both consummatory and motivational aspects of female sexuality in rats, the major difference being that the motivational components of the behavior have a higher threshold and longer latency before their manifestation (Glaser et al. 1983). Thus, the influence of progesterone on the consummatory (lordosis behavior) and motivational (proceptive behaviors) aspects of female sexuality in the rat seems to be the same. The situation is, however, quite different in other species; particularly those in which one can dissociate the hormonal mechanisms controlling receptive behaviors (willingness to accept the mounting male) from those controlling proceptive behaviors (solicitations to arouse the sexual interest of the male), e.g., in the rhesus monkey (Everitt et al. 1981). Also, it may be extremely rewarding to monitor motivational and consummatory aspects separately in studies on the neurochemical control of female sexual behavior (Hansen et al. 1981a, b).

Here we review published data on estrogen-progesterone effects on the sexual behavior of intact and ovariectomized rats, with no distinction being made between receptivity (lordosis) and proceptivity. Ever since the effect of progesterone on sexual behavior in the rat was first demonstrated (Boling and Blandau 1939), it has been well known that progesterone does not act independently but in synergy with estrogen. Thus, the behavioral effects of progesterone cannot be analyzed separately from those of estrogen. The characteristics of the estrogen stimulus necessary for induction of lordosis will therefore be discussed here. The possible role of progesterone in the termination of estrous behavior and the effect of progesterone on the estrogen conditioning process will also be discussed. In addition, the site and mechanism of progesterone action will be considered briefly.

3 Induction of Sexual Behavior in Intact Female Rats

3.1 Effect of Estradiol

The behavioral sex cycle of the female rat can easily be monitored by daily manual stimulation (Zucker 1967a; Södersten and Hansen 1977). Several hundred rats can be tested within an hour and the probability of unwanted pseudopregnancies or behavioral diestri is minimized. When the animals have shown two consecutive 4-day behavioral cycles, the probability that they will continue showing regular 4-day cycles is 100% (Södersten and Hansen 1977). The onset and termination of sexual behavior was studied in 16 rats by manual stimulation at hourly intervals, commencing 3 h before lights off and continuing for 20 h on the day of the expected behavioral estrus. This day was arbitrarily called day 4 and the other days 1, 2, and 3. The rats gradually became receptive starting 1 h before lights off; all rats showed sexual receptivity 4 h after lights off; and the animals ceased showing the behavior 17 to 20 h after the initiation of testing (Mean $\pm$ S.E.M. duration of receptivity: 13.3 ± 1.1 h). All subsequent tests were therefore performed 4 h after lights off, since normally at this time all rats were receptive. All estrogen injections were given 24 or 48 h before behavioral testing since, at the time, it was assumed that a 24 h periodicity might control the behavioral sex cycle in the rat, similar to the periodicity reported in the hamster (Alleva et al. 1971). Injection of estradiol benzoate (EB) on any day of the cycle, with the exception of day 2, was followed by display of sexual behavior 24 h later (Table 1) (Södersten and Hansen 1977).

Ovariectomy at the time of the EB injection blocked the behavioral response 24 h later, and exogenous administration of progesterone (P) 6 h before testing restored the display of sexual behavior in all rats. The behavior could also be induced in animals treated with EB on day 2 of the cycle, by P treatment 6 h before testing (Södersten and Hansen 1977). Thus, EB-induced sexual receptivity in intact rats requires the presence of the ovaries. The estrogen treatment probably stimulates the release of pituitary LH, which acts on recently formed, short-lived corpora lutea to stimulate P secretion, which then induces the behavior (Barraclough et al. 1971; Roos et al. 1973; Södersten and Hansen 1977). These results

Table 1. Induction of sexual receptivity by estradiol benzoate (EB) in groups of five rats with 4-day estrous cycles

Day of EB injection	Dose of EB (μg)	No. of rats receptive (%) 24 h after EB injection
1	0.25	0
1	0.5	60
1	1.0	100
1	2.0	100
2	1.0	0
2	2.0	0
2	4.0	0
2	6.0	0
2	100.0	20
3	0.25	100
3	0.5	100
3	1.0	100
3	2.0	100
4	0.25	0
4	0.5	20
4	1.0	100
4	2.0	100

showed that sexual receptivity can be induced by small amounts of exogenous EB given on any day of the estrous cycle and that P is required for the behavioral response. In no stage of the cycle was the behavioral response inhibited. Goy et al. (1966) had predicted that in species lacking a functional corpus luteum, such as the rat, inhibition of behavioral estrogen-P sensitivity during the estrous cycle should not be expected.

During the rat estrous cycle, serum P levels are elevated only during a relatively brief time on the day of behavioral estrus (Butcher et al. 1974; Södersten and Eneroth 1981 a) and during day 2 of the cycle (Butcher et al. 1974). Table 1 shows that the behavioral response to a low dose of EB (0.5 μg) given on day 4, i.e., at the time when the rats were receptive and had been exposed to high endogenous P levels, was reduced in comparison to the response to an EB injection given on day 1. A higher (but still relatively low) dose of EB (1.0 μg) eliminated the difference in EB sensitivity between day 4 and day 1 (Table 1 and Fig. 1). The

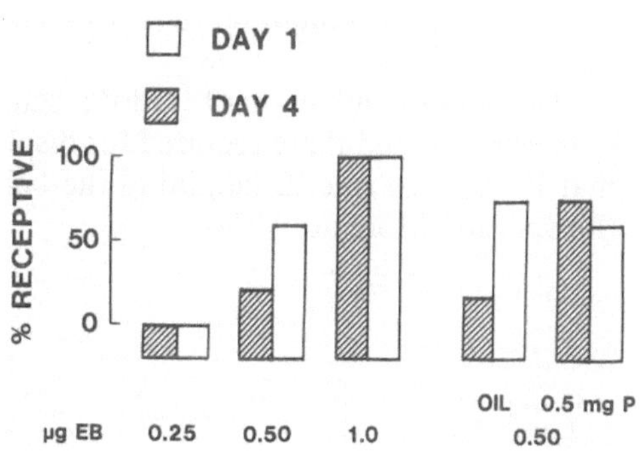

Fig. 1. Percent intact, cycling rats showing lordosis 24 h after treatment with estradiol benzoate (EB) on days 1 or 4 of the estrous cycle. Progesterone (P) (0.5 mg) or oil was injected 6 h before testing in rats given 0.5 μg EB (*right*)

Table 2. Effects of progesterone (P) on the display of sexual receptivity in rats with 4-day estrous cycles

No. rats	Treatment			No. of rats receptive (%) day 4
	mg P on day 2	day 3	day 4	
6	0	–	–	100
7	0.5	–	–	100
7	1.0	–	–	85.7
11	2.0	–	–	9.1
9	4.0	–	–	0
10	2.0	Vehicle	Vehicle	10
13	2.0	2.0 μg EB	Vehicle	23.1
10	2.0	2.0 μg EB	0.5 mg P	100

difference in behavioral EB sensitivity between day 4 and day 1 rats was also eliminated by injecting P on the day of testing (Fig. 1) (Powers and Zucker 1969).

These results show that there exists no stage of the rat estrous cycle in which behavioral estrogen sensitivity is reduced. This is in contrast to the guinea pig, where behavioral responses to EB and P are prevented immediately after the display of behavioral estrus (Goy et al. 1966) when serum P levels are low (Feder et al. 1968), and during the ensuing luteal phase (Goy et al. 1966).

Thus, endogenous P secretion during the estrous cycle does not inhibit behavioral EB sensitivity in rats. However, exogenous supply of P can prevent ovulation (Everett 1961) and, similarly, P treatment of intact rats can delay the occurrence of sexual receptivity by one or two days (Zucker 1967a). It has been suggested that such experimentally induced increased P levels somehow interfere with the subsequent conditioning action of estrogen (Zucker 1967a). Table 2 shows that injection of 2 mg P on day 2 of the cycle prevents the display of sexual behavior on day 4, and a subsequent injection of 2 μg EB on day 3 fails to restore the behavioral response on day 4. However, an injection of 0.5 mg P before testing on day 4 produces lordosis responses in all P + EB-treated rats. Obviously, P treatment in the middle of the rat estrous cycle does not prevent subsequent exogenous estrogen treatment from properly preparing the animal to show sexual behavior on the usual day. Most probably, P treatment of intact rats causes a 1 or 2 day delay in the display of sexual behavior and ovulation by delaying, not by preventing or inhibiting, the hormonal events preceding lordosis and ovulation.

The data on induction of lordosis behavior in cyclic female rats suggest that both estrogen and P are required for display of the behavior (Powers 1970), and that P plays no role in inhibiting the behavior under physiological conditions (Powers and Moreines 1976).

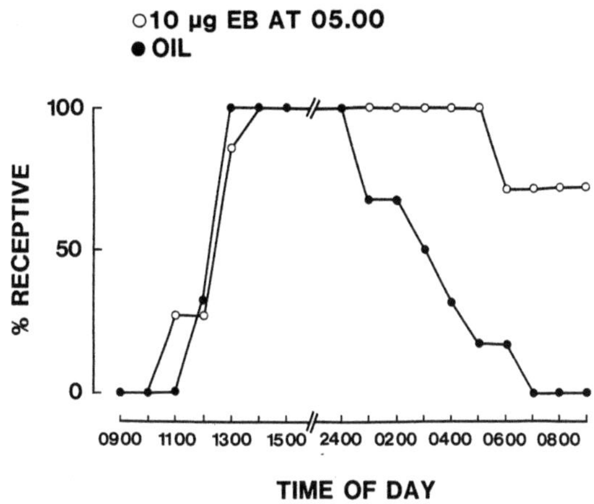

Fig. 2. Percent intact, cycling rats showing lordosis in tests at hourly intervals. The rats were injected with 10 µg estradiol benzoate (EB) or oil at 05.00 h on the day of behavioral estrus

3.2 Influence of Estradiol and Progesterone on the Duration of Sexual Receptivity

The influence of exogenously administered EB upon the onset and duration of sexual receptivity is shown in Fig. 2. Injection of 10 µg EB at 05.00 h on day 4 did not affect the time of onset of the behavior, but greatly prolonged the duration of the behavioral response. (Södersten and Hansen 1977). Since induction of lordosis by EB during the estrous cycle was shown to be dependent on ovarian P secretion, which in turn is mediated by pituitary LH secretion (see above), it seems likely that the failure of EB to advance the onset of the behavioral response was due to the coupling of the LH release apparatus to the LD cycle (Everett and Sawyer 1950), i.e., the behavioral response to EB cannot occur before LH release has been activated and this must await the appropriate circadian signal, which occurs daily at the same time of day (Legan et al. 1975). Lower doses of EB (0.5–2.0 µg) affected neither the onset nor the duration of the behavior, although these EB doses are well above the threshold for induction of lordosis in cyclic rats (Södersten and Hansen 1977). However, rats injected with 2 µg EB on day 4 were sexually receptive 24 h later and again after 48 h, but not 72 h after the injection (Södersten and Hansen 1977). Yet this dose of EB did not affect the duration of the period of receptivity; these EB-treated rats became receptive 24 h after the injection and then, after this period of receptivity had ceased, the animals entered another period of behavioral estrus 48 h after the EB injection. It seems possible that the circadian pattern of LH release, which can be activated by exogenous EB treatment and which is reflected in daily discharges of large amounts of LH by the pituitary gland (Legan and Karsch 1975) had been activated in these EB-treated rats, ovarian P secretion occured in response to the LH release, and as a consequence the animals went into repeated periods of sexual heat. For this to occur, estrogen levels must be continously raised (Legan and Karsch 1975), and normally this is not the case. Also, endogenous P secretion can prevent the expression of daily LH surges during the rat estrous cycle (Banks and Freeman 1978). EB treatment produces prolonged elevation of serum estradiol concentration in rats (Tapper et al. 1974) and it seems possible that these levels exceed those produced by the ovaries under physiological conditions.

These results suggest the possible involvement of estradiol in determination of the duration of behavioral heat (Moreines and Powers 1977), but in view of the high doses of EB required, it is likely that other factors in addition to endogenous estrogen levels contribute to the duration of the period of behavioral estrus.

In rats with 5 day estrous cycles, it was shown that P treatment could advance the display of sexual receptivity by 1 day (Zucker 1967a) indicating that the animals are actually prepared to show the behavior 1 day earlier than it is normally displayed, i.e., the rats were appropriately conditioned with estrogen 24 h before the normal time of behavioral estrus. A possible reason for the absence of an early behavioral response is that estrogen levels, although sufficiently high for behavioral responses, may have been too low to activate the circadian signal necessary for the LH release apparatus. The activation of this signal is, however, not an absolute requirement for induction of lordosis behavior, since cyclic rats are capable of showing sexual receptivity before the time at which they normally do so. Thus, P treatment early on day 4 of the cycle advances the onset of the behavioral response by several hours (Barfield and Lisk 1970; Södersten and Hansen 1977). Figure 3 shows that injection of 2 mg P at 06.00 h on day 4 of the cycle advanced the onset of sexual receptivity by about 4 h.

In an early experiment, P-induced lordosis behavior in estrogen-treated ovariectomized guinea pigs was reported to last for a shorter period of time than lordosis behavior induced by estrogen alone (Boling et al. 1938). The refractoriness of guinea pigs to EB during the luteal phase of the estrous cycle (Goy et al. 1966) had indicated that P may exert inhibitory effects on sexual behavior. It was therefore suggested that P, in addition to stimulating the behavior, might activate a mechanism which eventually terminates the period of receptivity (Morin and Feder 1973). In the cyclic rat, however, inhibition by P did not appear to play a role in the termination of sexual behavior. Figure 3 shows that advancement of the onset of behavioral estrus by P treatment did not affect the time at which the

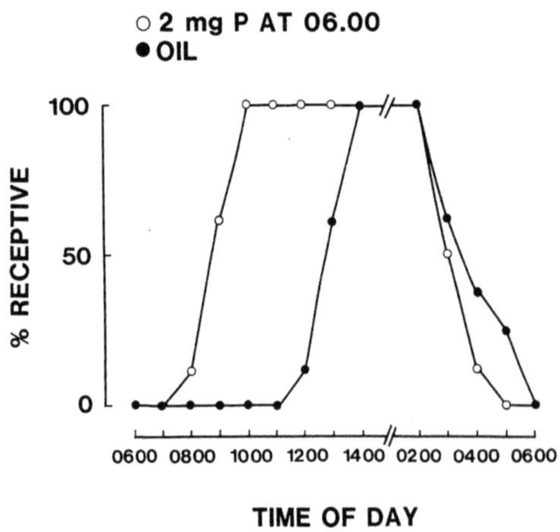

Fig. 3. Percent intact, cycling rats showing lordosis in tests at hourly intervals. The rats were injected with 2.0 mg progesterone (P) or oil at 05.00 h on the day of behavioral estrus

behavior terminated, as might have been expected if P, at the time it facilitates sexual behavior, also activates a mechanism which later terminates the behavior. Also, administration of 2 mg P every third hour to sexually receptive rats did not affect the time at which the behavior ceased (Södersten and Hansen 1977). These results are in line with those of several investigations on intact and ovariectomized guinea pigs, which report failure to abbreviate the period of estrous behavior by repeated P injections to receptive animals (Goldfoot and Goy 1970; Joslyn et al. 1971; Morin and Feder 1973; Wallen and Thornton 1979). Therefore, while P is clearly essential for induction of sexual behavior, it seems unlikely that it is related to the mechanisms which eventually terminate the period of receptivity.

3.3 Blockade of Estrogen-Induced Sexual Receptivity in Pregnant and Pseudopregnant Rats

Goy et al. (1966) predicted that in species lacking a functional luteal phase of the estrous cycle, such as the rat, inhibition of behavioral estrogen sensitivity during the cycle would be unlikely. This prediction turned out to be correct (see Sect. 3.2). In view of the relatively higher behavioral P sensitivity of guinea pigs compared to rats, Powers and Zucker (1969) argued that P inhibition of sexual behavior might occur in rats only during reproductive states when serum P levels are continuously elevated, i.e., during an induced functional luteal phase such as pregnancy and pseudopregnancy. As predicted, pregnant and pseudopregnant rats were refractory to the behavioral effects of 2 µg EB. However, neither pregnant nor pseudopregnant rats were totally insensitive to EB, since a higher, yet reasonably low, dose of EB (6 µg) induced relatively high LQs (about 80). Interestingly, females in the second half of pregnancy did not respond to any of the estrogen treatments, whereas early-pregnant rats did. It was suggested that the high endogenous P levels of late-pregnant rats interfered with the estrogen conditioning process. The inhibition of sexual behavior in pregnant rats has recently been confirmed, although a different explanation has been offered (Baum et al. 1979; De Greef et al. 1981; De Greef and Merkx 1982).

4 Induction of Sexual Behavior in Ovariectomized Rats

4.1 Estradiol-Progesterone Interactions in the Induction of Sexual Receptivity

Ovariectomized (OVX) rats, in contrast to intact rats, do not show sexual receptivity 24 h after an injection of EB (Södersten and Hansen 1977, 1979); the EB has to be given at an earlier time to stimulate the behavior. On the hypothesis that a circadian control mechanism exists which influences the behavioral responses of female rats to ovarian hormones (Södersten et al. 1981 a), EB was administered 48 h before testing of OVX rats.

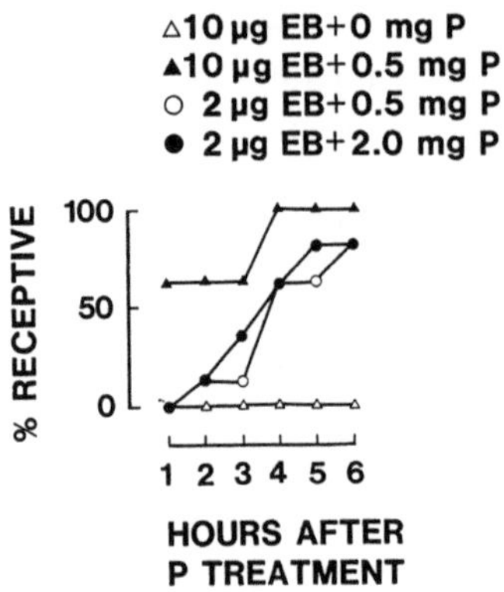

Fig. 4. Percent ovariectomized rats showing lordosis in tests at hourly intervals. The rats were injected with 2 or 10 µg estradiol benzoate (EB) 48 h before testing, in combination with 0, 0.5 or 2.0 mg progesterone (P) given 6 h before testing

Figure 4 shows that, whereas injection of 10 µg EB failed to induce lordosis behavior 48 h later, injection of a lower dose of EB (2 µg) induced the behavior in most rats if followed 42 h later by an injection of 0.5 mg P. Neither a high dose of P (2 mg) nor administration of supplementary EB (6 µg) 2 h before the P injection reduced the latency to the behavioral response in animals given 2 µg EB; but the response to the lower P dose (0.5 mg) appeared significantly earlier in animals primed with a high dose of EB (10 µg) than in animals given the low EB dose (2 µg). Thus, OVX rats must be exposed to estrogen for a period of 24–48 h before they are appropriately prepared to show sexual behavior in response to P treatment. The estrogen dose, rather than the P dose, determines the latency of the behavioral response.

The difference between intact and OVX rats in the duration of the estrogen stimulus needed for induction of lordosis may be more apparent than real. It was shown that the short-latency response, i.e., 24 h, of intact rats to 2 µg EB was lost 24 h after OVX and could not be restored by increasing the dose of EB (20 µg) (Table 3) (Södersten and Hansen 1979). This suggested that in intact rats ovarian secretions before the injection of EB contributed to the behavioral response to the EB injection. These ovarian secretions are probably estrogens, since antiestrogen treatment of intact rats during the 24 h before OVX and EB injection, through subcutaneous insertion and subsequent removal of antiestrogen (CI-628) implants, prevented the short latency behavioral response to the EB (Table 3). Also, subcutaneous insertion of constant release implants filled with oil solutions of estradiol in low concentration (15.0 µg/ml), which in themselves were below the threshold concentration needed for induction of the behavior in OVX rats, restored the short latency response to an injection of 2 µg EB in OVX rats (Table 3). The results suggest that cyclic rats may be exposed to small amounts of estrogens during the major part of the estrous cycle and that only a relatively brief acute increase in serum estrogen concentrations is needed for induction of behavioral P sensitivity and the subsequent display of sexual receptivity. Evidence that these low estrogen levels may be behaviorally relevant was provided by the observation that injection of high doses of P (4 mg) induced high LQs (74.3 ± 4.8) in seven out of ten intact rats on days 1 and 3 of the cycle, in the absence of prior estrogen treatment, i.e., at times of the cycle when rats normally do not show sexual behav-

Table 3. Lordosis quotients in groups of five acutely ovariectomized (OVX) rats[a]

	Treatment or test		
0 h	24 h	LQ tests 48 h	LQ tests 72 h
OVX+2 μg EB	LQ=100.0±0	–	–
OVX	2 μg EB	14.0± 8.7	–
OVX	20 μg EB	20.0±15.5	–
OVX	2 μg EB	–	100.0± 0
CI-628 implant	OVX+2 μg EB CI-628 removed	18.0±11.1[b]	76.0±16.6
Control implant	OVX+2 μg EB control removed	94.0± 6.0	90.0± 5.5
OVX+15.0 μg/ml E_2 implant	–	8.0± 8.0[b]	–
OVX+15.0 μg/ml E_2 implant	2 μg EB	92.0± 3.7	–

[a] Six h before each LQ test, 0.5 mg progesterone was given
[b] $P<0.05$ compared to the appropriate control (Mann-Whitney U test)

ior (Södersten and Hansen 1979). OVX rats do not respond to P unless pretreated with EB. Evidence that a brief period of increased estrogen stimulation is sufficient for induction of sexual behavior in cyclic rats was provided by the observation that subcutaneous insertion and subsequent removal, 6 h later, of a constant-release implant filled with crystalline estradiol, was found to induce high LQs (80–100) in day 1 cyclic rats which were not treated with P before testing (Hansen et al. 1978).

4.2 Repeated Activation of Sexual Receptivity by Progesterone Stimulation at Short Time Intervals in Estrogen-Treated Rats

The strength of the estrogen and P stimuli given to OVX rats can be kept constant by using constant-release implants filled with crystalline estradiol or P. Serum levels of estradiol or P are elevated within 1 h of subcutaneous placement of these implants, remain constant for the duration of the implantation period, and decline to low levels within a few hours of implant removal (Södersten and Eneroth 1981 b; Södersten et al. 1981 b). Subcutaneously placed estradiol implants, prepared from Silastic Medical Grade Tubing no. 602-285 and only 1 mm in length, provide a very potent stimulus for induction of sexual receptivity in OVX rats (Ydstebo and Södersten 1977). Using implants filled with crystalline P (prepared from the same size Silastic Tubing) in estradiol-implanted OVX rats, it was found that P implants only 2 mm in length were sufficient for induction of high LQs (Fig. 5) (Hansen and Södersten 1978). Twenty-four h after testing, all estradiol-implanted rats showed the same LQs independent of the continuous presence of previously implanted 2 or 10 mm P implants (Fig. 5). Placement of another P implant 6 h before the second test stimulated maximal lordosis behavior in all rats,

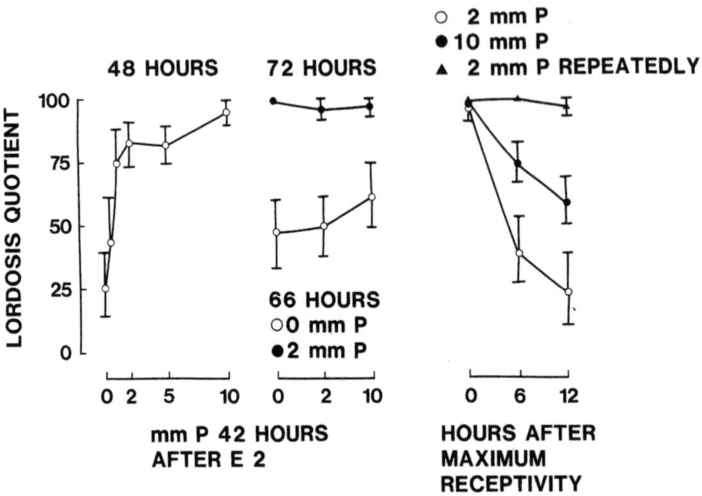

Fig. 5. Mean ± S.E.M. lordosis quotients in ovariectomized rats given 1 mm estradiol (E_2)-filled implants at 0 h. The rats were given progesterone (P)-filled implants of various lengths 42 and 66 h after E_2 implantation, and were tested twice 6 h after P implantation (*left, center*). The lordosis quotients of rats made receptive by 1 mm E_2 implants in combination with 2 or 10 mm P implants and tested at 6 h intervals are shown to the *right*

despite the presence of the previous P implants (Fig. 5). These results show that a constant P stimulus facilitates lordosis behavior in estrogen-pretreated rats only once, that the behavioral effect of the P treatment wanes within 24 h, and that another period of receptivity can be activated with a low level of P stimulation despite the presence of a constant amount of P in excess of that needed for induction of the behavior. This suggests a "triggering" function of P in the control of feminine behavior. Such a function is to be expected, considering the secretory pattern of P during the estrous cycle (Fig. 7).

Although the duration of the period of sexual receptivity was not affected by exogenous P treatment in cyclic rats (see above) the duration of the period of sexual responding in OVX rats is dependent upon the amount of P given to the animals. Figure 5 shows that the decline of sexual behavior in animals made receptive by estradiol and P implants depended on the length of the P implant. Animals which were receptive after treatment with 1 mm estradiol implants in combination with 10 mm P implants showed higher LQs in behavioral tests conducted at 6 h intervals than animals given a combination of 1 mm estradiol implants and 2 mm P implants. Interestingly, decline of the behavior in both groups was independent of the continued presence of either the estradiol or P implants. Thus, removal of either of these implants after the first test, i.e., after maximum sexual receptivity was manifest, had no effect on the decline of the behavior (Hansen and Södersten 1978).

These results showed that induction of sexual behavior by a constant amount of P and continued presence of this P stimulus does not prevent subsequent P treatments from inducing another period of sexual receptivity 24 h later. Thus, doses of P which are sufficiently high or even above the threshold for stimulation of lordosis behavior have no inhibitory effect upon the subsequent behavioral responses to renewed P stimulation. Figure 5 shows that the behavior can be reactivated at very short time intervals. Thus, OVX rats made receptive by insertion of 1 mm estradiol implants and 2 mm P implants were maximally receptive in subsequent tests given at 6 h intervals, if another 2 mm P implant was given immediately after each test. Obviously, the P-sensitive neural system which controls the display of sexual behavior to the pattern, rather than the absolute strength, of the applied P stimulus. Since serum levels of P produced by subcutaneous P implants are proportional to the length of the implants (Södersten and Eneroth 1981 b), it must be assumed that the rats repeatedly given 2 mm P implants ($2 \times 3 = 6$ mm) had lower serum P levels than the rats given a single 10 mm P implant.

4.3 Inhibitory Effects of Progesterone on Sexual Receptivity

The luteal phase inhibition of sexual receptivity in the guinea pig (Goy et al. 1966) suggested that P might inhibit estrogen-P–induced lordosis in this species. Zucker (1966, 1968), Zucker and Goy (1968), and Wallen et al. (1975) clearly defined the conditions under which P might interfere with the induction of lordosis by EB and P. It was shown that after the period of receptivity induced by EB and P, another period of sexual behavior was unlikely to occur in response to either renewed P treatment or to renewed EB + P treatment. Thus, P was said to exert "biphasic" effects on sexual behavior in guinea pigs, first facilitating and then inhibiting the behavior. Additionally, it was suggested that P might exert antiestrogenic effects on behavior by acting as a competitive estrogen inhibitor in the brain. Initial attempts to demonstrate similar behavioral effects of P in the rat were unsuccessful (Whalen and Nakayama 1965; Zucker 1967 b; Edwards et al. 1968; Nadler 1970), but evidence was subsequently presented that more or less identical effects to those described in the guinea pig could be demonstrated in rats, the only difference being that rats are apparently somewhat less sensitive to the inhibitory actions of P than guinea pigs (Powers and Moreines 1976; Blaustein and Wade 1977 a, b; Marrone et al. 1977). Two types of inhibitory effects were proposed originally by Powers and Moreines (1976): concurrent inhibition and sequential inhibition. In the concurrent paradigm, P, if administered at the same time as the EB, will prevent subsequent P treatment from inducing sexual behavior. In the sequential paradigm, EB is administered to an OVX animal and P is then given to facilitate the behavior and, after the period of facilitation is over, inhibit the behavior (for details see Feder and Marrone 1977). Sequential inhibition by P is equivalent to what Zucker (1968) referred to a the biphasic effect of P. In analogy with the guinea pig literature it was suggested that P might interfere with the neural effects of estrogen, i.e., act as an estrogen antagonist in the brain (Blaustein and Wade 1977 a, b). The physiological importance of the demonstration of inhibitory effects of P on female rat sexual behavior was not immediately apparent, since no inhibitory effects of P had been demonstrated during the estrous cycle (see above). Likewise, the proposed antiestrogenic effect of P was doubtful, as there were several reports showing that other estrogen-dependent be-

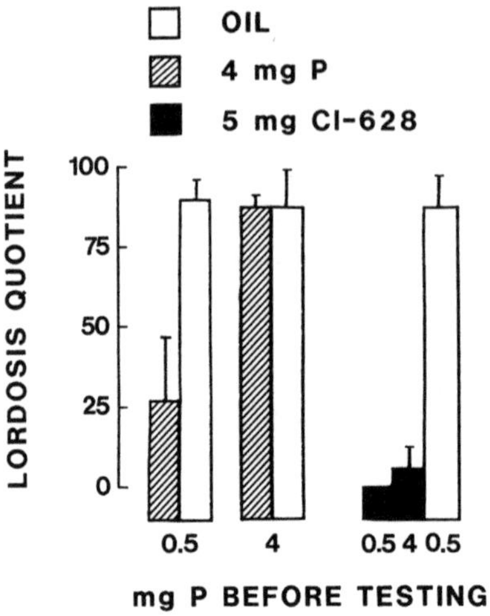

Fig. 6. Mean ± S.E.M. lordosis quotients in ovariectomized rats treated with 2 µg estradiol benzoate in combination with oil, 4 mg progesterone (P), or 5 mg of the antiestrogen CI-628 and tested 48 h later. P (0.5 or 4 mg) was administered 6 h before testing

haviors in female rats, such as male-like mounting behavior (Södersten 1974) and body weight regulatory behaviors (Tarttelin and Gorski (1973), were uninfluenced by P treatment. Reduced frequencies of mounting behavior and alterations in body weight regulatory behaviors during functional luteal phases were suggested to be caused by reduced ovarian estrogen secretion rather than by inhibitory effects of the elevated levels of serum P (Södersten 1974; Tarttelin and Gorski 1973). Moreover, it was found that inhibition of sexual receptivity in OVX rats, in both the concurrent and sequential paradigms, could be overcome by increasing the amount of P given prior to testing (Hansen and Södersten 1979). Thus, although OVX female rats given 4 mg P and 2 µg EB simultaneously were unable to respond to a second injection of 0.5 mg P 42 h later, they readily showed lordosis if the second P dose was increased to 4 mg (Fig. 6). However, animals given 5 mg of the antiestrogen CI-628 concurrently with the EB did not display lordosis after a subsequent injection of 4 mg P (Fig. 6). Also, it has been long known that a moderate level of sexual receptivity can be induced in OVX rats by high doses of EB in the absence of P stimulation (Davidson et al. 1968 a, b; Södersten 1976). Such EB-induced lordosis behavior was unaffected by concurrent P treatment, but was inhibited by concurrent antiestrogen treatment (Hansen and Södersten 1979).

These results showed that concurrent P treatment, in contrast to concurrent antiestrogen treatment, does not interfere with the conditioning action of estradiol. This casts doubt on the notion that P can act as a behavioral antiestrogen. Similar findings were reported by Rodriguez-Sierra and Davis (1978). Further evidence that P does not antagonize the behavioral effect of estrogen was provided by the demonstration that sexual behavior could readily be induced in pseudopregnant rats by reasonable doses of EB, provided the dose of P injected prior to testing was sufficiently high. Thus, injection of 2 µg EB in day 5 pseudopregnant rats, followed by an injection of 50 mg P 42 h later, induced LQs of 100 in four out of six rats (Hansen and Södersten 1979). Day 15 pregnant rats, however, were resistant to this hormone treatment. P treatment of OVX rats, therefore, does not exert an antiestrogenic effect on sexual behavior nor does it make the animals refractory to renewed P stimulation; rather, the threshold for renewed P stimulation is increased. This effect of P is best referred to P desensitization induced by prior P facilitation (Blaustein 1982a, b; Blaustein et al. 1982; Södersten and Eneroth 1982).

5 Sexual Receptivity and Ovarian Hormones Under Physiological Conditions

5.1 Estradiol and Progesterone Serum Levels and Sexual Receptivity in Intact and Ovariectomized Rats

Measurements of serum levels of estradiol and P were performed in groups of intact, regularly cycling rats decapitated during various phases of the cycle; the measurements were then related to the onset, duration, and termination of sexual receptivity (Södersten and Eneroth 1981a). Figure 7 shows that, what one characteristically sees in peripheral blood under these sampling conditions is a great variation in serum estradiol levels. Contrary to some previous reports (Brown-Grant et al. 1970; Butcher et al. 1974; Shaikh and Shaikh 1975; Smith et al. 1975), no smooth proestrous increase in serum estradiol concentrations was

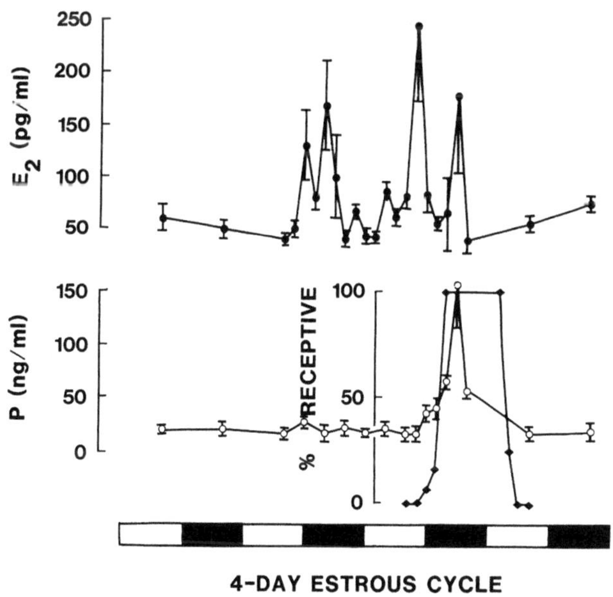

Fig. 7. Mean ± S.E.M. serum levels of estradiol (E_2) and progesterone (P) in groups of intact, cycling rats decapitated at various times of the estrous cycle. A group of rats were tested for sexual receptivity at hourly intervals. The light : darkness cycle is indicated by the black and white horizontal bars

observed, rather, in agreement with a recent report (Nequin et al. 1979), serum estradiol concentrations fluctuated markedly, particularly during the approximately 35 h preceding the onset of behavioral estrus.

In close parallel with increasing serum P concentrations, the rats gradually became sexually receptive and all rats were receptive before serum P levels reached their maximal concentrations (Fig. 7). These results suggest that under physiological conditions a relatively small increase in serum P is required for induction of sexual receptivity, that P acts with a very short latency to induce the behavior, and that P is secreted in considerably higher quantities than needed for induction of the behavior. It is known that P can activate lordosis behavior in OVX EB-treated rats within 15 min if administered intravenously (Lisk 1960) or intracerebrally (Ross et al. 1971).

Serum levels of estradiol and P in OVX rats were kept constant by subcutaneous insertion of constant-release implants filled with oil solutions of various concentrations of estradiol (Silastic Tubing, no. 602-285, 30 mm in length) or crystalline P (Silastic Tubing, no. 601-335, of various lengths). Serum levels of estradiol and P produced with such implants vary within the physiological range (Fig. 8).

Figure 9 shows that although implantation of estradiol in OVX rats stimulated the display of lordosis in the absence of P treatment, the LQs shown by these animals never reached the maximum, i.e., 100, shown by intact cycling rats tested at the same time of day. Yet the implants (480 μg estradiol/ml) produced levels of serum estradiol which were continuously elevated, for a 48–72 h period, to higher levels than those observed in intact rats. This is direct evidence that stimulation with estradiol alone is insufficient for activation of normal-intensity sexual receptivity under physiological conditions.

Facilitation of sexual behavior, in OVX rats primed with doses of estradiol which were within or below the physiological range, occurred with P implants which produced serum P levels of <50 ng/ml (Figs. 8 and 9). Animals given a higher, yet still relatively low, estradiol dose (160 μg estradiol/ml) required less P than animals primed with a very low dose of estradiol (30 μg estradiol/ml). These P levels are far below the normal levels observed during the estrous cycle (about 150 ng/ml, Södersten and Eneroth 1981 a). This is direct evidence that much less P is needed for induction of sexual behavior in OVX rats than is secreted by the preovulatory ovary. Since during the estrous cycle all rats were sexually receptive

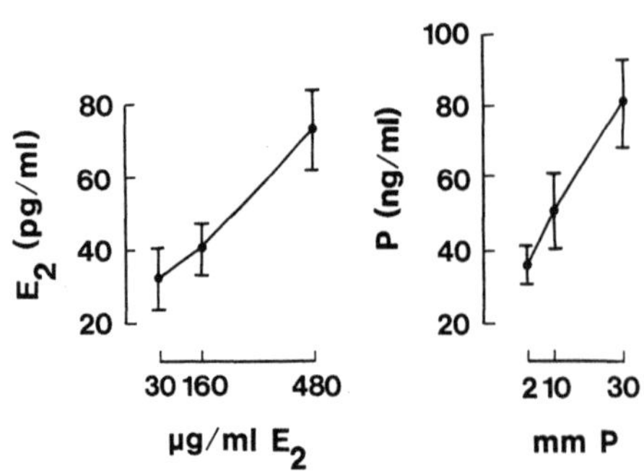

Fig. 8. Mean ± S.E.M. serum levels of estradiol (E_2) and progesterone (P) in ovariectomized rats given implants filled with various concentrations of E_2, or P-filled implants of various lengths

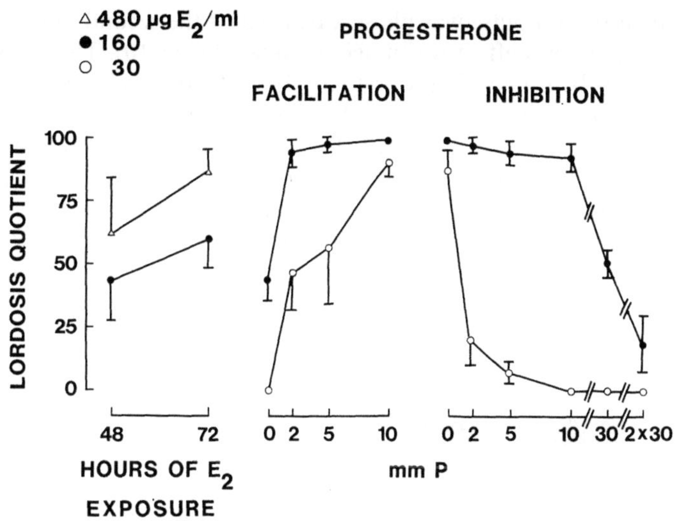

Fig. 9. Mean ± S.E.M. lordosis quotients in ovariectomized rats given implants filled with various concentrations of estradiol (E_2). Facilitation: progesterone (P)-filled implants of various lengths were given 6 h before testing. Inhibition: P-filled implants of various lengths were given concurrently with the E_2 and 10 mm P implants were given 6 h before testing

before the time at which maximal serum P levels were found (Fig. 7), these data suggest that the very high concentrations of serum P which are secreted by the preovulatory rat ovary must be related to some other function than induction of behavioral estrus; e.g., preparing the uterus for blastocyst implantation (Psychoyos 1976).

Concurrent implantation of estradiol and P showed that inhibition or desensitization to subsequent P treatment, 42 h later, could be observed when the hormones were administered in physiological concentrations (Fig. 9); animals primed with 160 µg estradiol/ml required considerably more P concurrently with the estradiol for subsequent P desensitization than animals given 30 µg estradiol/ml (see also Blaustein and Feder 1979).

5.2 Behavioral Effects of Pulsatile Administration of Estradiol

The characteristics of the estradiol stimulus necessary for induction of lordosis behavior in OVX rats were studied, again through subcutaneous insertion and subsequent removal of estradiol-filled constant-release implants (Södersten et al. 1981 b). Figure 10 shows that stimulation with estradiol must start no less than 32 h before behavioral testing (with 0.5 mg P given 6 h before the test) for induction of high LQs in OVX rats. The estradiol stimulus could be discontinued 20 h after its initation without loss of behavioral effect. More interesting, two 4-h pulses of estradiol stimulation, given in the beginning and end of the 20 h period, could replace an uninterrupted period of estradiol stimulation. If, however, the first estradiol pulse was phase advanced or the second pulse was phase delayed, the behavioral effect was much attenuated (Fig. 10). These findings, and similar findings by others (Clark and Roy 1983; Harris and Gorski 1978; Parsons et al. 1981, 1982) are of importance for understanding the mechanisms of action of the behavioral effect of estradiol. Two hypotheses were proposed to explain the estrogen conditioning process: a "triggering" versus a "maintenance" hypothesis. In the first, estrogen has to be present only for a brief time to trigger all of the biochemical events necessary for induction of behavioral P sensitivity and the subsequent display of sexual receptivity. According to the alternative hypothesis,

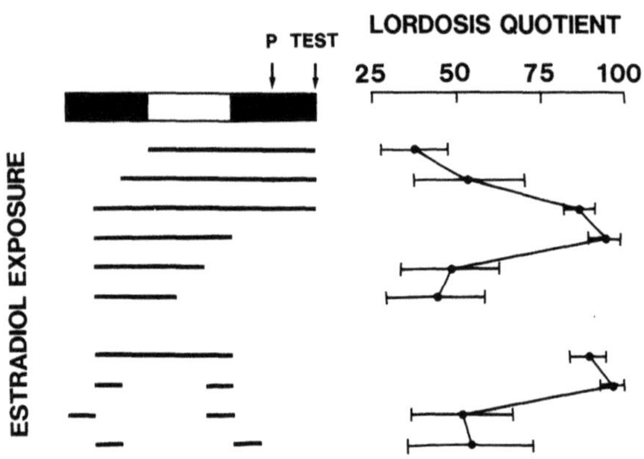

Fig. 10. Mean ± S.E.M. lordosis quotients in ovariectomized rats exposed to 1 mm estradiol-filled implants during various times (*horizontal black lines*) before testing. Progesterone (P) (0.5 mg) was given 6 h before testing (*arrow*). The light: darkness cycle is indicated by the black and white horizontal bars

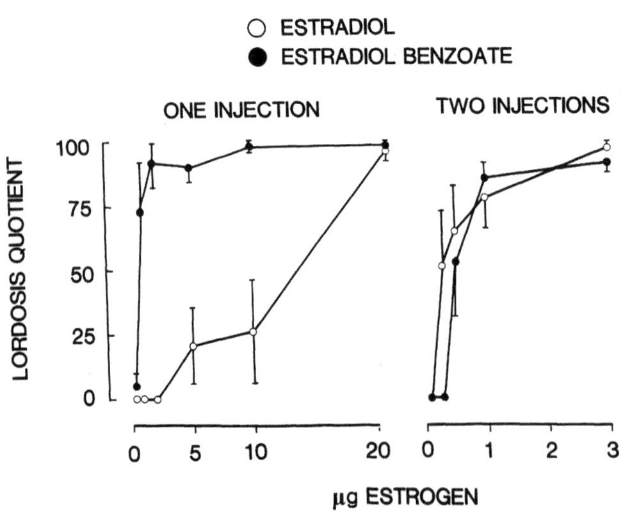

Fig. 11. Mean ± S.E.M. lordosis quotients in ovariectomized rats given one or two injections of various doses of estradiol benzoate or estradiol

prolonged estrogen stimulation is required before P can facilitate the behavior (reviewed by Feder et al. 1979). An argument for the "maintenance" hypothesis was provided by the demonstration that long-acting forms of estradiol, e.g., EB, are considerably more potent than short-acting forms, i.e., non-esterified estradiol, in inducing the behavior. Figure 11 shows that a much higher dose of estradiol is needed for induction of high LQs in OVX rats than the EB dose required for the same behavioral response. However, if two injections are given according to the time schedule outlined above (Fig. 10), the difference in potency is reversed (Fig. 11). It appears that two brief pulses of estradiol are sufficient, although not an absolute requirement, for induction of lordosis in OVX rats (see also Pfaff and McEwen 1983).

It was noticed that serum levels of estradiol fluctuated markedly in intact cyclic rats decapitated during various stages of the estrous cycle (Fig. 7), and it was suggested that this might be a reflection of a pulsatile pattern of estradiol secretion by the preovulatory rat ovary, as has been reported in other species (Baird 1978). Subsequently, this was shown to be the case. Sequential blood samples, taken at hourly intervals from freely moving proestrous rats, showed that estradiol is secreted episodically (Fig. 12). Subcutaneous injections of low doses of es-

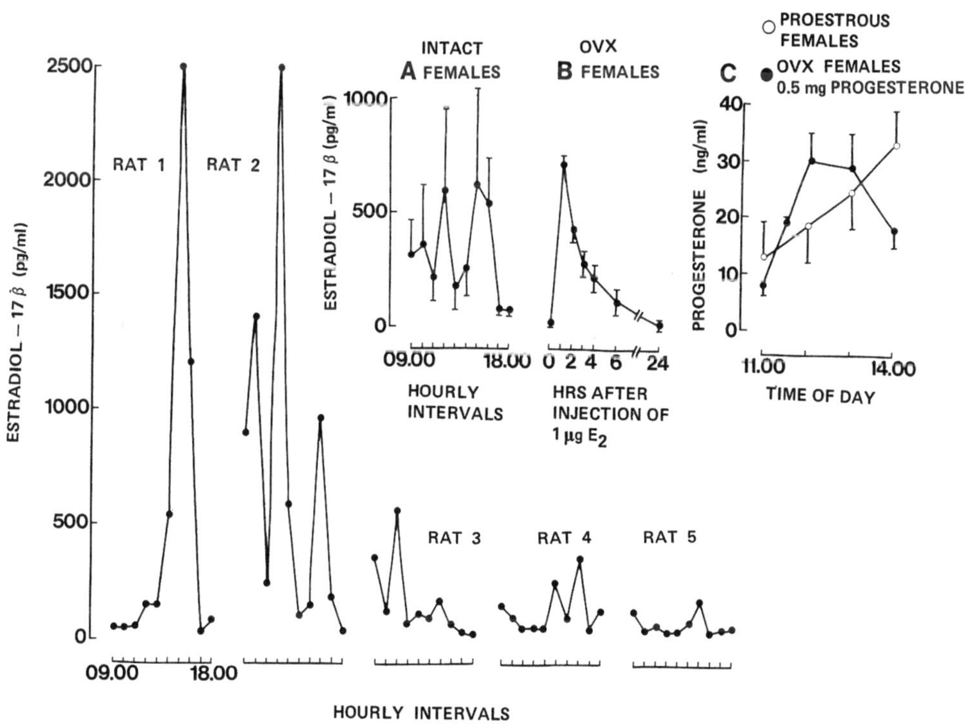

Fig. 12. Mean ± S.E.M. serum levels of estradiol-17β (E_2) in 5 intact, cycling rats bled at hourly intervals between 09.00 and 18.00 h on the day before behavioral estrus (*A*). E_2 levels of individual rats are shown in the lower part of the figure. *B* shows E_2 levels in ovariectomized (OVX) rats injected with 1 µg E_2 and *C* shows progesterone levels in OVX rats injected with 0.5 mg progesterone and in intact proestrous rats

tradiol (1 μg) according to the schedule outlined above, followed by an injection of 0.5 mg P, induced high levels of sexual receptivity and produced serum levels of estradiol and P which mimicked those seen under physiological conditions (Fig. 12) (Södersten et al. 1983 a). Additionally, serum LH levels fluctuate in a pulsatile fashion in cycling rats (Gallo 1981; Södersten and Eneroth 1982). Thus, in the rat as in other mammals (Bäckström et al. 1982; Baird 1978), pituitary and ovarian hormone secretion occurs in a pulsatile rather than steady state manner. This is a reflection of the pulsatile secretion of gonadotropin releasing hormone by the brain (Knobil et al. 1980; Levine et al. 1982). In view of these findings it seems highly likely that the behavioral effect of the pulsatile schedule of estradiol administration represents a physiologically relevant phenomenon.

5.3 Progesterone Is Not a Behavioral Antiestrogen

The determination of the precise characteristics of the estrogen stimulus sufficient for induction of lordosis (see Sect. 5.2) allowed a more detailed description of the way in which P interacts with estrogen to cause facilitation of the behavior and subsequent desensitization to further P stimulation (Södersten and Eneroth 1981 b). OVX animals were given 5 mm estradiol implants and tested precisely 32 h later. Figure 13 shows that P (a 30 mm implant) could be administered any time during the 32 h between the onset of the estrogen stimulation and behavioral testing: facilitation of the behavior occurred in all groups provided that P was present during the last 6 h before testing. This is direct evidence that P does not affect the estrogen conditioning process in rats. Others have reached the same conclusion (Rodriguez-Sierra and Davis 1978).

In a test 48 h after the combined estradiol-P treatment, the animals displayed the same LQs as animals given estradiol without concurrent P (Figs. 5 and 13). After the second test, P treatment (subcutaneous injection of 0.1 mg P) of animals bearing both estradiol and P implants failed to stimulate the display of lordosis. Thus, P acts first to facilitate sexual receptivity in estradiol-treated OVX rats independently of the time of administration and, after the period of facilitation is over, desensitizes the system to renewed P stimulation. These observations would

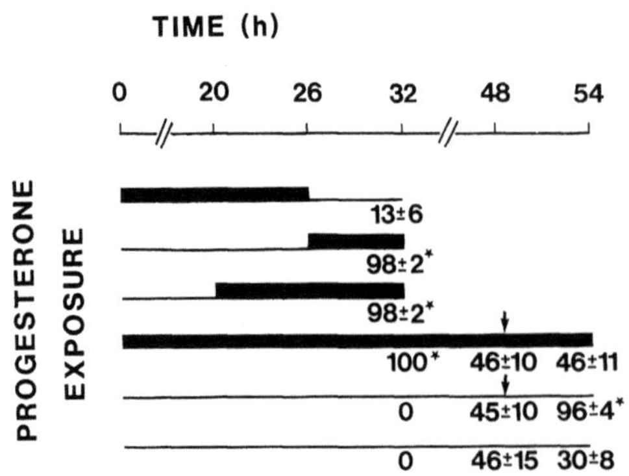

Fig. 13. Mean ± S.E.M. lordosis quotients in ovariectomized rats given 5 mm estradiol (E_2)-filled implants at 0 hours. The rats were exposed to 30 mm progesterone (P)-filled implants during various times between the onset of E_2 treatment and testing (*black horizontal bars*).P (0.1 mg) was injected immediately after testing at 48 h (*arrow*). *$P < 0.05$ compared to the appropriate control group (Mann-Whitney *V* test)

Table 4. Sexual receptivity induced by two injections of 1 μg estradiol on day 5 and 6 in pseudopregnant rats; and serum levels of estradiol and progesterone in pseudopregnant rats not treated with estradiol

No. of rats	% showing lordosis	LQ	Estradiol (pg/ml)		Progesterone (ng/ml)	
			day 5	day 6	day 5	day 6
5	80	80±12	16±2	17±4	84±23	96±8

seem to undermine the usefulness of the concept of concurrent inhibition by P. It seems that no such inhibitory effect of P exists.

Since P does not antagonize the estradiol conditioning process in the rat, and since P can be administered at any time in relation to a 32 h estradiol priming period, it should be possible to induce high levels of sexual behavior in pseudopregnant rats by the appropriate estrogen treatment, without P treatment before testing. This was shown to be possible (Södersten and Eneroth 1981 b). Two injections of 1 µg estradiol, 32 and 16 h before testing, induced high LQs in day 6 pseudopregnant rats (Table 4). Serum levels of estradiol were low and P levels were high in rats decapitated on days 5 and 6 of pseudopregnancy; the times when the estradiol was administered to the animals used in the behavioral tests (Table 4). Also, day 10 pregnant rats ($n=5$) showed lordosis with the same estrogen treatment (80% responded, LQ 57.5±8.5).

The possibility that P desensitization may be relevant in understanding the hormonal events controlling the display of lordosis in cycling rats was examined by determining the minimum duration of P exposure necessary to produce desensitization. OVX animals were given a 120 µg estradiol/ml implant which produced low concentrations of serum estradiol (38±9 pg/ml); the animals were then given a low dose of P (0.1 mg) 42 h later and tested for lordosis 48 h after the initiation of estrogen treatment. The rats were exposed to P implants (30 mm in length) during various phases of the estradiol conditioning period. Figure 14 shows that significant desensitization occurred only if the P implants remained for at least a 32-h period. The serum P levels produced by the P implants were 55–70 ng/ml. Since serum P levels in intact regularly cycling females are not elevated above this level for more than 4 h during the proestrus P surge (Fig. 7, Södersten and Eneroth 1981 a), these results rule out the possibility that P desensitization plays a role in the control of the behavioral sex cycle of the female rat of our strain.

A difference between OVX and intact sexually receptive rats in their behavioral response to repeated P treatment should be apparent: whereas it was possible to maintain high levels of sexual receptivity in OVX rats by repeated stimulation with short P implants (Fig. 5) (Hansen and Södersten 1978), it was not possible to prolong the period of sexual receptivity in intact rats be repeated injections of even large doses (2 mg) of P (Södersten and Hansen 1977). It is possible that short-term desensitization to P may occur in intact rats, since the amplitude of the proestrous P surge is very high (Fig. 7) (Södersten and Eneroth 1981 a). However, desensitization in the intact animal is unlikely to play a physiological

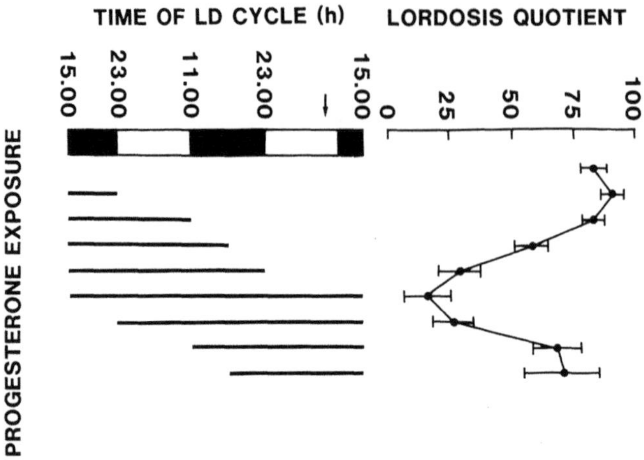

Fig. 14. Mean ± S.E.M. lordosis quotients in ovariectomized rats given implants filled with 120 μg estradiol/ml and exposed to progesterone (P)-filled implants during various times between the onset of the estradiol treatment and behavioral testing (*black horizontal lines*) P (0.1 mg) was given 6 h before testing (*arrow*). The light: darkness (LD) cycle is indicated by the black and white horizontal bars

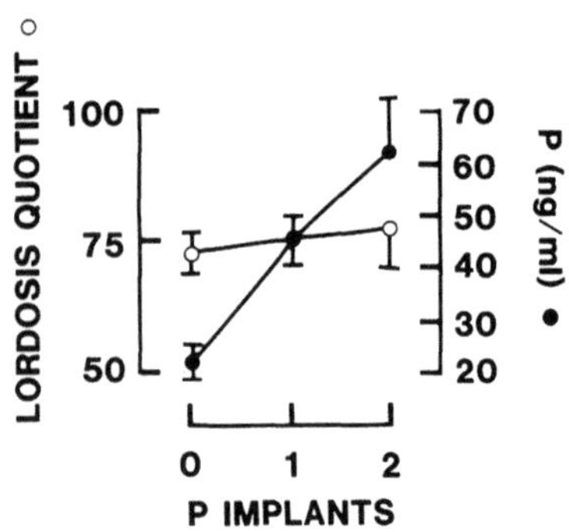

Fig. 15. Mean ± S.E.M. lordosis quotients and serum progesterone (P) levels in ovariectomized rats given 5 mm estradiol-filled implants in combination with various numbers of 30 mm P-filled implants. The values are overall means of three tests given 2, 3, and 4 days after implantation

role, since no P is secreted during the cycle after the proestrous P surge (Södersten and Eneroth 1981 a), and after induction of a functional luteal phase P levels rise only slowly, several days after the P surge (Smith et al. 1975). In addition, it was quite possible to stimulate another period of sexual receptivity in intact rats, in the absence of exogenous estrogen treatment, by injection of P on the day after the display of sexual receptivity (see above, Södersten and Hansen 1979). A relatively high dose of P (2 mg) was needed for induction of the behavioral response in these day 1 cycling rats; this was probably due to the low levels of serum estradiol during this phase of the cycle rather than to the preceding proestrous P surge, since supply of exogenous EB (2 μg) on the day before restored the behavioral P sensitivity of day 1 rats (Södersten and Hansen 1977).

A direct test of the proposed antiestrogenic effect of P was made using 5 mm estradiol implants in combination with different numbers of 30 mm P implants in OVX rats, and testing the animals 2, 3, and 4 days later, without further P treatment before testing. Figure 15 shows that despite the fact that serum P levels varied depending upon the number of P implants present, the LQs (calculated as overall means ± S.E.M. of the 3 tests) were identical in all groups. Clearly, P is not a behavioral antiestrogen. The same conclusion has been reached by others (Baum et al. 1979).

5.4 Effects of Other Hormones on the Induction of Sexual Receptivity

Although most of the experimental work on the endocrine control of female sexuality in rats has been concentrated on the roles of estradiol and P; the influence of other hormones of hypothalamic, pituitary, and gonadal origin should be considered. For example, the preovulatory LH surge in female rats is preceded by a surge of LH-releasing hormone (LH-RH) secretion by the brain (Sarkar et al. 1976) and LH-RH treatment of estrogen-treated OVX rats facilitates lordosis behavior (Pfaff 1973; Sakuma and Pfaff 1980). Thus, although the proestrous LH-RH surge may not be an absolute requirement for display of lordosis behavior (Harlan and Gorski 1978) it may play a role under physiological conditions.

5α-Reduced androgens have been reported to inhibit behavioral estrogen sensitivity in OVX rats (Baum and Vreeburg 1976), perhaps by acting on the lateral septum (Tobet and Baum 1982). This inhibitory effect of androgens may be physiologically relevant, since antiandrogens and 5α-reductase inhibitors can prolong the duration of the period of sexual receptivity in cycling rats (Erskine 1983), and there is a surge of ovarian androgen secretion coinciding with the phase of the estrous cycle when the rats are sexually receptive (Gay and Tomacari 1974). Androgen inhibition of sexual receptivity may be one of the mechanisms responsible for termination of the behavior under physiological conditions.

There is a rhythmic variation in the levels of arg-vasopressin (AVP) in the cerebrospinal fluid (but not blood) during the LD cycle, with maximum levels during the L phase (Reppert et al. 1981). During this phase sexual receptivity is inhibited in OVX estradiol-treated rats (Hansen et al. 1979). Intracerebroventricular AVP injections (but not peripheral ones) inhibit the behavior of sexually receptive rats (Södersten et al. 1983 b). Thus, another factor which may be involved in terminating the period of sexual receptivity under physiological conditions is the secretion of AVP by the brain (Södersten et al. 1983 b).

During the estrous cycle there is a massive surge of prolactin (PRL) secretion, triggered by the previous increase in ovarian estradiol secretion (Smith et al. 1975). Recently, the possibility was investigated that the proestrous PRL surge might participate in the induction of sexual receptivity in rats (Södersten et al. 1983 c). Dopamine is the main inhibitor of hypophyseal PRL secretion (Neill 1980), and serum PRL levels can be conveniently raised in rats by using the dopamine-receptor antagonist domperidone, which does not cross the blood-brain barrier (thus, the possibility of a direct effect on dopamine neurons is avoided) (Costall et al. 1979). Figure 16 shows that injection of 2.5 mg domperidone facilitated the induction of lordosis in EB-treated (2 μg) OVX rats within 3 h of intraperitoneal administration, with a time course which was comparable to that of P (0.5 mg). Both domperidone treatment and P treatment produced a marked increase in serum PRL levels, and a slight increase in serum P levels was

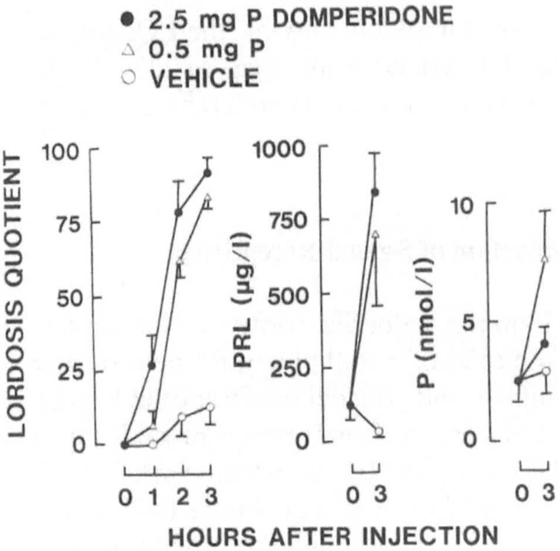

Fig. 16. Mean ± S.E.M. lordosis quotients in ovariectomized rats treated with 2 μg estradiol benzoate and tested 32 h later. The rats received vehicle, 0.5 mg progesterone (P) or 2.5 mg domperidone 3 h before testing. Serum levels of prolactin (PRL) and P are also shown

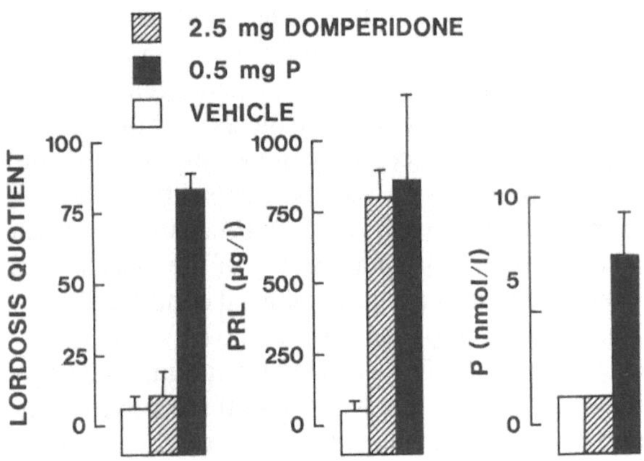

Fig. 17. Mean ± S.E.M. lordosis quotients in ovariectomized and adrenalectomized rats treated with 2 μg estradiol benzoate and tested 32 h later. The rats received vehicle, 0.5 mg progesterone (P), or 2.5 mg domperidone 3 h before testing. Serum levels of prolactin (PRL) and P are also shown

noticed after the domperidone treatment. P, of course, also elevated serum P levels. However, the stimulatory effect of domperidone on sexual behavior and serum P levels (but not the effect on serum PRL levels), was eliminated by adrenalectomy (Fig. 17). Adrenalectomized, OVX, EB-treated rats readily displayed lordosis in response to P treatment (Fig. 17). Thus, although an acute increase in serum PRL levels cannot in itself stimulate sexual behavior, PRL can do so indirectly by activating the secretion of an adrenal factor, probably P, which in turn facilitates lordosis.

Estrogen stimulates PRL secretion in rats (Neill 1980) and it is possible that EB-induced PRL secretion, although incapable of independent action, might act in synergy with the subsequently administered P to stimulate the induction of lordosis. This seems unlikely, as daily treatment with the dopamine-receptor agonist bromocriptine, in doses which inhibit PRL secretion (0.5 mg) (Hansen et al. 1983), did not affect the induction of lordosis by EB and P in OVX rats (Södersten et al. 1983c). Obviously, the release of large amounts of PRL by the pituitary gland does not facilitate lordosis. Also, high endogenous PRL levels maintained for 6 days by daily domperidone treatment (2.5 mg/day or 2 × 2.5 mg/day) had no effect on the subsequent response to EB and P (Södersten et al. 1983c). It seems, therefore, that high levels of serum PRL exert no effect on the sexual behavior of female rats. However, a role of PRL in female rat behavior should not be excluded. Prolactin-like immunoreactivity has been demonstrated in nerve cells in rat hypothalamus (Fuxe et al. 1977) and recent evidence indicates that PRL infused into the dorsal midbrain can potentiate lordosis in OVX EB-treated rats (Harlan et al. 1983). However, another recent study (Dudley et al. 1982) showed that intracerebroventricular PRL infusions inhibited the behavior of sexually receptive rats. The conditions under which intracerebrally administered PRL can either facilitate or inhibit female rat sexual behavior must therefore be more precisely defined.

6 Inhibition of Sexual Behavior in Lactating Rats

Hyperprolactinemia is associated with reduced reproductive capacity, including impaired sexual behavior, in man and experimental animals (e.g., Bailey and Herbert 1982; Evans et al. 1982; Perryman and Thorner 1981). In the female rat, lactation is the reproductive state which is characterized by physiological hyperprolactinemia and high serum P levels (Hansen et al. 1983). However, no inhibitory effects of either high PRL or high P levels on behavioral estrogen sensitivity could be demonstrated in female rats (see Sect. 5.4). Yet a preliminary observation by Numan (1974) indicated that lactating rats did not show lordosis in response to exogenous EB treatment. Injections of various doses of EB in day 5 lactating rats nursing 6 pups, and behavioral testing 32 h later with and without various doses of P given 3 h before testing, showed that sexual behavior is markedly inhibited

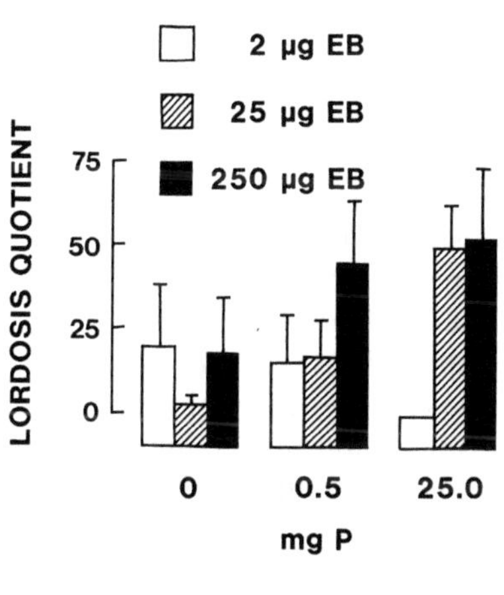

Fig. 18. Mean ± S.E.M. lordosis quotients in day 6 lactating rats nursing 6 pups and treated with various doses of estradiol benzoate (EB) and progesterone (P) 32 and 3 h before testing

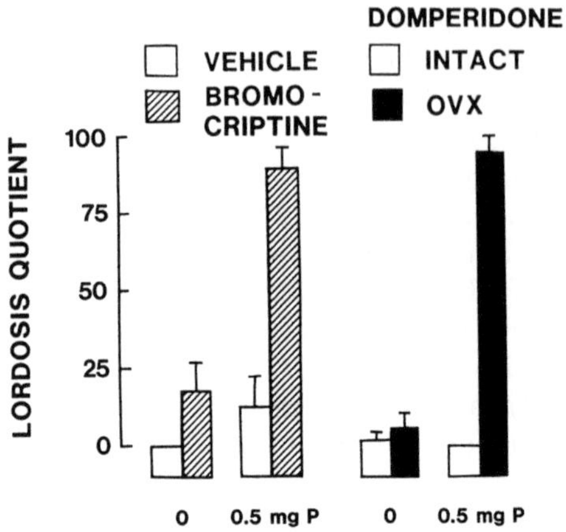

Fig. 19. Mean ± S.E.M. lordosis quotients in day 6 lactating rats nursing 6 pups and treated with vehicle or bromocriptine (0.5 mg/day), or with domperidone (2 × 2.5 mg/day) after pup removal and ovariectomy (OVX) or sham ovariectomy (intact). Estradiol benzoate (2 µg) and 0.5 mg P were injected 32 and 3 h before testing

in lactating rats (Fig. 18) (Södersten et al. 1983d). Injections of even extremely high doses of EB (250 µg) and P (25 mg) induced LQs of only 50 (Fig. 18).

Removal of the pups from lactating rats 48 h before the EB injection restored the capacity to show sexual behavior. The inhibitory effect of the presence of pups during the 48 h interval before the EB injection was proportional to the number of pups present. OVX of lactating rats permitted the induction of sexual receptivity by EB and P, despite the presence of suckling pups. Maternal and sexual behavior can obviously be displayed at the same time by lactating rats, but sexual behavior is normally suppressed during lactation. This inhibition is caused not by the suckling stimulus per se but by some of its consequences, e.g., the release of PRL. Thus, lowering serum PRL levels by treating lactating rats with bromocriptine restored the capacity for showhing lordosis after EB + P treatment (Fig. 19). Moreover, PRL acts on the ovaries of the mother; for pup removal in combination with maintainenance of endogenous PRL levels by daily domperidone treatment resulted in retention of the state of refractoriness to the behavioral effects of the EB + P treatment only if the ovaries remained in situ (Fig. 19). Thus, during lactation, suckling-induced PRL release causes the secretion of an ovarian factor, which subsequently acts alone or in concert with another factor to inhibit behavioral estrogen sensitivity. OVX and hormone replacement treatment with P, or androgens such as testosterone and 5α-dihydrotestosterone, identified P as the ovarian factor. Lowering endogenous PRL levels in OVX P-treated lactating rats by daily bromocriptine treatment restored behavioral estrogen sensitivity despite the presence of suckling pups; pup removal in combination with artificial maintenance of high PRL levels by daily domperidone treatment in OVX P-treated lactators prevented the induction of behavioral estrus by EB + P. These results indicate that during lactation, sexual receptivity is inhibited due to a combined action of PRL and P (Södersten et al. 1983d). Interestingly, although neither PRL nor P is in itself capable of affecting behavioral estrogen sensitivity, they can act in synergy to prevent estrogen from properly conditioning an animal for the display of sexual behavior in response to subsequent P treatment. A similar explanation for the inhibition of sexual receptivity in late-pregnant rats has recently been offered (De Greef et al. 1981).

Inhibition of sexual behavior during lactation may be relevant for efficient reproduction in rats. Although rats have the capacity for showhing maternal and sexual behavior simultaneously (see above) high serum PRL and P levels prevent the expression of sexual behavior during lactation. This may be important for appropriate maternal care, which in turn is essential for the nutritional and psychological development of the neonatal rat pup, which needs almost full-time maternal attention due to its relative immaturity at birth. Activation of the sexual motivation system during lactation might cause the mother rat to leave the nest to receive copulatory stimulation from males, thereby neglecting the needs of the developing young. Thus, PRL might, in synergy with P, act to cause a motivational switch in the lactating rat, directing all of her behavioral attention to the newborn pups.

7 Site and Mechanism of Progesterone Action in Control of Sexual Receptivity

It has been long known that intracerebral P administration can facilitate sexual behavior in rodents (Kent and Liberman 1949). One mechanism through which estradiol induces lordosis in response to P treatment is by increasing the number of P receptors in the brain (Parsons and Pfaff, this volume; Pfaff and McEwen 1983). There is a conspicuous correlation between the increase in hypthalamic cytosol P receptor concentrations and the increase in behavioral P sensitivity with time after EB treatment (Parsons and Pfaff, this volume; Pfaff and McEwen 1983). Also, estradiol induces P receptors in many brain areas, for example the arcuate nucleus, the medial preoptic area, the ventromedial hypothalamic nucleus (VMH), and the periventricular region of the anterior hypothalamus (Rainbow et al. 1982). With this background it is hardly surprising that behavioral effects of intracerebral P implants have been reported from a variety of sites. To name a few: facilitation of lordosis has been reported with implants in the VMH-arcuate region (Morin and Feder 1974a; Rubin and Barfield 1983), the medial basal hypothalamus (Powers 1972; Rodriguez-Sierra and Komisaruk 1982), the mesencephalic reticular formation (Ross et al. 1971), the nucleus caudatus-putamen (Yanase and Gorski 1976a, b), and the habenula (Tennent et al. 1982). P desensitization of lordosis has been reported with implants in the substantia nigra (Morin and Feder 1974b) and the medial basal hypothalamus (Marrone et al. 1979). However, general agreement has recently been reached that the VMH is the most effective site, both for estradiol conditioning and P facilitation of lordosis (Davis et al. 1979, 1982; Pfaff and McEwen 1983). The research strategy used in the experiments implicating the critical importance of the VMH in the neuroendocrine control of female rat sexual behavior has been very successful in dissecting out small pieces of hypothalamic tissue, which is extremely sensitive to estradiol and P, and, hence, has been very useful in the selection of an appropriate tissue sample for studying the biochemical actions of estradiol and P on nerve cells. However, at the same time it has somewhat distracted our attention from the study of other potential sites of hormone action. It seems unlikely that all of the estradiol–P–LH-RH–AVP–androgen–PRL interactions take place at only a few discrete hypothalamic sites. Some of these interactions might not even take place in the brain. Sites of sex hormone action in behavioral control outside the brain have been suggested (Beach 1948) and found (Kow and Pfaff 1973).

It has been known for some 20 years that alterations in brain monoamine neurotransmission can influence the display of sexual receptivity (Crowley and Zemlan 1981). However, only recently has an analysis been carried out of the relationship between the behavioral effects of the ovarian hormones, i.e., estradiol and P, and the amines. This is surprising in view of the extensive knowledge of the rather precise endocrine requirements for display of sexual behavior in female rodents. Nock et al. (1981) and Nock and Feder (1981) have recently suggested that cerebral monoamines may modulate tissue steroid sensitivity via an alteration in receptor availability. Thus, the noradrenaline system must remain intact to permit estradiol to induce P receptors. Interference with noradrenaline neurotransmission, by treatment with a dopamine-β-hydroxylase inhibitor, significantly reduced the concentration of hypothalamic P receptors found after EB treatment (Nock et al. 1981). Central neural sites at which such steroid-monoamine interactions may occur are presently being explored (Luine and Fischette 1982). The results of these studies are awaited with great interest.

8 Summary

Estradiol is secreted in a pulsatile manner under physiological conditions, and a pulsatile schedule of estradiol administration effectively prepares an OVX rat to respond to the subsequent administration of P by showing sexual receptivity. P stimulation is required for display of the behavior under physiological circumstances, and P activates the behavior by a "trigger" mechanism. If continuously available in supraphysiological concentrations, P first facilitates induction of the receptive behavior and then desensitizes the animal to renewed P stimulation. P desensitization has not yet been shown to be a physiologically relevant phenomenon. Termination of sexual behavior in receptive rats may be caused by ovarian androgen secretion and by hypothalamic peptide secretion. During lactation, sexual receptivity is not displayed by female rats. Refractoriness to the behavioral effects of estrogen and P during lactation is the result of a synergistic action of P and PRL. P can exert behavioral effects from a variety of neural sites, possibly by interacting with central aminergic systems.

Acknowledgements. This review is dedicated to Professor David Magnusson on the occasion of his 60th birthday. The work was performed in collaboration with Drs. S. Hansen and P. Eneroth. I thank Mr. T. Arvidson of Sandoz AG for donating bromocriptine and Dr. W. van Bever of Janssen Pharmaceutica for donating domperidone; also the Journal of Endocrinology Ltd and the Williams & Williams & Wilkins Company for permission to use previously published material. The work was supported by the Swedish Council for Research in the Humanities and Social Sciences, The Bank of Sweden Tercentenary Foundation, and the Karolinska Institute.

References

Adler NT, Bell D (1969) Constant estrus in rats: vaginal, reflexive and behavioral changes. Physiol Behav 4:151–153
Alleva JJ, Waleski MV, Alleva FR (1971) A biological clock controlling the estrous cycle of the hamster. Endocrinology 88:1368–1379
Bäckström CT, McNeilly AS, Leask RM, Baird DT (1982) Pulsatile secretion of LH, FSH, prolactin, oestradiol and progesterone during the human menstrual cycle. Clin Endocrinol 17:29–35
Bailey DJ, Herbert J (1982) Impaired copulatory behavior of male rats with hyperprolactinaemia induced by domperidone or pituitary grafts. Neuroendocrinology 35:186–193
Baird DT (1978) Pulsatile secretion of LH and ovarian estradiol during the follicular phase of the sheep estrous cycle. Biol Reprod 18:359–364
Banks JA, Freeman ME (1978) The temporal requirement of progesterone on proestrus for extinction of the estrogen-induced daily signal controlling luteinizing hormone release in the rat. Endocrinology 102:426–432
Barfield MA, Lisk RD (1970) Advancement of behavioral estrus by subcutaneous injections of progesterone in the four day cyclic rat. Endocrinology 87:1096–1098
Barraclough CA, Collu R, Massa R, Martini L (1971) Temporal interrelationships between plasma LH, ovarian secretion rates of progesterone and peripheral plasma progestin concentrations in the rat: effects of nembutal and exogenous gonadotropins. Endocrinology 88:359–371
Baum MJ, Vreeburg JTM (1976) Differential effects of the anti-estrogen MER-25 and of three 5α-reduced androgens on mounting and lordosis behavior in the rat. Horm Behav 7:87–104
Baum MJ, De Greef WJ, Kloet A, Schretlen PJM (1979) Evidence that a factor besides progesterone, prolactin, or plasma-estradiol-binding protein inhibits estrogen-induced sexual receptivity in pregnant rats. J Comp Physiol Psychol 93:278–294
Beach FA (1942) Importance of progesterone to induction of sexual receptivity in spayed female rats. Proc Soc Exp Biol Med 31:369–371
Beach FA (1948) Hormones and behavior. Hoeber, New York
Beach FA (1976) Sexual attractivity, proceptivity, and receptivity in female mammals. Horm Behav 7:105–138
Blaustein JD (1982a) Progesterone in high doses may overcome progesterone's desensitization effects on lordosis by translocation of hypothalamic progestin receptors. Horm Behav 16:175–190
Blaustein JD (1982b) Alteration of sensitivity to progesterone facilitation of lordosis in guinea pigs by modulation of hypothalamic progestin receptors. Brain Res 243:287–300
Blaustein JD, Feder HH (1979) Progesterone at plasma levels lower than those of mid-pregnancy decreases sexual behavior in ovariectomized rats. Physiol Behav 23:1099–1104
Blaustein JD, Wade GN (1977a) Concurrent inhibition of sexual behavior, but not (^{3}H)estradiol uptake, by progesterone in female rats. J Comp Physiol Psychol 91:742–751
Blaustein JD, Wade GN (1977b) Sequential inhibition of sexual behavior by progesterone in female rats: comparison with a synthetic anti-estrogen. J Comp Physiol Psychol 91:752:760
Blaustein JD, Brown TJ, Reading DS (1982) Failure of protein synthesis inhibition to block progesterone desensitization of lordosis in female rats. Physiol Behav 29:475–481
Boling JL, Blandau RJ (1969) The estrogen-progesterone induction of mating responses in the spayed female rat. Endocrinology 25:359–364
Boling JL, Young WC, Dempsey EW (1938) Miscellanous experiments on the estrogen-progesterone induction of heat in the spayed guinea-pig. Endocrinology 23:182–187
Brown-Grant K, Exley D, Naftolin F (1970) Peripheral plasma oestradiol and luteinizing hormone concentrations during the oestrous cycle of the rat. J Endocrinol 48:295–296

Butcher RL, Collins WE, Fugo NW (1974) Plasma concentrations of LH, FSH, prolactin, progesterone and estradiol-17β throughout the 4-day estrous cycle of the rat. Endocrinology 94:1704–1708
Clark AS, Roy EJ (1983) Behavioral and cellular responses to pulses of low doses of estradiol-17β. Physiol Behav 30:561–566
Costall B, Fortuna DH, Naylor RJ (1979) Neuropharmacological studies on the neurological potential of domperidone (R 33812). J Phar Pharmacol 31:344–347
Crowley WR, Zemlan FD (1981) The neurochemical control of mating behavior. In: Adler NT (ed) Neuroendocrinology of reproduction. Plenum, New York, pp 451–484
Davidson JM, Rogers CH, Smith ER, Bloch GJ (1968 a) Stimulation of female sex behavior in adrenalectomized rats with estrogen alone. Endocrinology 82:193–195
Davidson JM, Smith ER, Rogers CH, Bloch GJ (1968 b) Relative threshold of behavioral and somatic responses to estrogen. Physiol Behav 3:227–229
Davis PG, McEwen BS, Pfaff DW (1979) Localized behavioral effects of tritiated estradiol in the ventromedial hypothalamus of female rats. Endocrinology 104:898–903
Davis PG, Krieger MS, Barfield RJ, McEwen BS, Pfaff DW (1982) The site of action of intrahypothalamic estrogen implants in feminine sexual behavior: an autoradiographic analysis. Endocrinology 111:1581–1586
De Greef WJ, Merkx JAM (1982) Receptivity, proceptivity and attractivity of pregnant and pseudopregnant rats. Behav Brain Res 4:203–208
De Greef WJ, Schenck PE, Vreeburg JTM, van der Vaart PDM, Baum MJ (1981) Evidence that a placental factor other than androsterone or dihydrotestosterone inhibits oestrogen-induced lordosis behavior in pregnant rats. J Endocrinol 89:13–23
Dempsey EW, Herz R, Young WC (1936) The experimental induction of oestrus (sexual receptivity) in the normal and ovariectomized guinea pig. Am J Physiol 116:201–209
Dudley CA, Jamieson TS, Moss RL (1982) Inhibition of lordosis behavior in the female rat by intraventricular infusion of prolactin and by chronic hyperprolactinemia. Endocrinology 110:677–679
Edwards DA, Pfeifle JK (1983) Hormonal control of receptivity, proceptivity and sexual motivation. Physiol Behav 30:437–443
Edwards DA, Whalen RE, Nadler RD (1968) Induction of estrus: estrogen-progesterone interactions. Physiol Behav 3:29–33
Erskine MS (1983) Effects of an antiandrogen and 5α-reductase inhibitor on estrus duration in the cyclic female rat. Physiol Behav 30:519–524
Evans WS, Cronin MJ, Thorner MO (1982) Hypogonadism in hyperprolactinemia: possible mechanisms. In: Ganong WF, Martini L (eds) Frontiers in neuroendocrinology, vol 7. Raven, New York, pp 77–122
Everett JW (1961) The mammalian female reproductive cycle and its controlling mechanisms. In: Young WC (ed) Sex and internal secretions, vol 2. Williams and Wilkins, Baltimore, pp 497–555
Everett JW, Sawyer CH (1950) A 24-hour periodicity in the "LH-release apparatus" of female rats, disclosed by barbiturate sedation. Endocrinology 47:198–218
Everitt BJ, Herbert J, Keverne EB, Martensz ND, Hansen S (1981) Hormones and sexual behavior in rhesus and talapoin monkes. In: Fuxe K, Gustafsson J-A, Wetterberg L (eds) Steroid hormone regulation of the brain. Pergamon, New York, pp 317–330
Feder HH, Marrone BL (1977) Progesterone: its role in the central nervous system as a facilitator and inhibitor of sexual behavior and gonadotropin release. Ann New York Acad Sci 286:331–354
Feder HH, Resko JA, Goy RW (1968) Progesterone concentrations in the arterial plasma of guinea pigs during the oestrous cycle. J Endocrinol 40:505–513
Feder HH, Landau IT, Walker WA (1979) Anatomical and biochemical substrates of the action of estrogens and antiestrogens on brain tissues that regulate female sex behavior of rodents. In: Beyer C (ed) Endocrine control of sexual behavior. Raven Press, New York, pp 317–340
Foster MA, Hisaw FL (1935) Experimental ovulation and the resultant pseudopregnancy in anoestrous cats. Anat Rec 62:75–93
Fuxe K, Hökfelt T, Eneroth P, Gustafsson J-A, Skett P (1977) Prolactin-like immunoreactivity: localization in nerve terminals of rat hypothalamus. Science 196:899–901

Gallo RV (1981) Pulsatile LH release during the ovulatory LH surge on proestrus in the rat. Biol Reprod 24:100–104

Gay VL, Tomacari RL (1974) Follicle-stimulating hormone secretion in the female rat: cyclic release is dependent on circulating androgen. Science 184:75–77

Glaser JH, Rubin BS, Barfield RJ (1983) Onset of the receptive and proceptive components of female sexual behavior in rats following intravenous administration of progesterone. Horm Behav 17:18–27

Goldfoot DA, Goy RW (1970) Abbreviation of behavioral estrus in guinea pigs by coital and vagino-cervical stimulation. J Comp Physiol Psychol 72:426–434

Goy RW, Phoenix CH, Young WC (1966) Inhibitory action of the corpus luteum on the hormonal induction of estrous behavior in the guinea pig. Gen Comp Endocrinol 6:267–275

Hansen S, Södersten P (1978) Effects of subcutaneous implants of progesterone on the induction and duration of sexual receptivity in ovariectomized rats. J Endocrinol 77:373–379

Hansen S, Södersten P (1979) Reversal of progesterone inhibition of sexual behaviour in ovariectomized rats by high doses of progesterone. J Endocrinol 80:381–388

Hansen S, Södersten P, Srebro B (1978) A daily rhythm in the behavioral sensitivity of the female rat to oestradiol. J Endocrinol 77:381–388

Hansen S, Södersten P, Eneroth P, Srebro B, Hole K (1979) A sexually dimorphic rhythm in oestradiol-activated lordosis behavior in the rat. J Endocrinol 83:267–274

Hansen S, Stanfield EJ, Everitt BJ (1981 a) The role of ventral bundle noradrenergic neurons in sensory components of sexual behaviors and coitus-induced pseudopregnancy. Nature 286:152–154

Hansen S, Stanfield EJ, Everitt BJ (1981 b) The effects of lesions of lateral tegmental noradrenergic neurons on components of sexual behavior and pseudopregnancy in female rats. Neuroscience 6:1105–1117

Hansen S, Södersten P, Eneroth (1983) Mechanisms regulating hormone release and the duration of dioestrus in the lactating rat. J Endocrinol 99:173–180

Harlan RE, Gorski RA (1978) Dissociation between release of luteinizing hormone and prolactin and induction of female sexual behavior in normal and androgenized female rats. Biol Reprod 19:439–446

Harlan RE, Shivers BD, Pfaff DW (1983) Midbrain microinfusions of prolactin increase the estrogen-dependent behavior, lordosis. Science 219:1451–1453

Harris J, Gorski J (1978) Evidence for a discontinuous requirement of estrogen in stimulation of deoxyribonucleic acid synthesis in the immature rat uterus. Endocrinology 102:240–245

Joslyn WD, Wallen K, Goy RW (1971) Cyclic changes in sexual responses to exogenous progesterone in female guinea pigs. Physiol Behav 7:915–917

Kent GC, Liberman MJ (1949) Induction of psychic estrus in the hamster with progesterone administration via the laterial brain ventricle. Endocrinology 45:29–32

Knobil E, Plant TM, Wildt L, Belchetz PE, Marshall G (1980) Control of rhesus monkey menstrual cycle: permissive role of hypothalamic gonadotropin-releasing hormone. Science 207:1371–1373

Kow L-M, Pfaff DW (1973) Effects of estrogen treatment on the size of the receptive field and response threshold of the pudendal nerve in the female rat. Neuroendocrinology 13:299–313

Legan SJ, Karsch (1975) A daily signal for the LH surge in the rat. Endocrinology 96:57–62

Legan SJ, Coon CA, Karsch FJ (1975) Role of estrogen as initiator of the daily LH surges in the ovariectomized rat. Endocrinology 96:50–56

Levine JE, Pau KF, Ramirez VD, Jackson GL (1982) Simultaneous measurement of luteinizing hormone-releasing hormone and luteinizing hormone release in unanesthetized ovariectomized sheep. Endrocrinology 111:1449–1455

Lisk RD (1960) A comparison of the effectiveness of intravenous, as opposed to subcutaneous, injection of progesterone for the induction of estrous behavior in the rat. Can J Biochem Physiol 38:1381–1383

Lodder J, Zeilmaker GH (1976) Role of the pelvic nerves in the postcopulatory abbreviation of behavioral estrus in female rats. J Comp Physiol Psychol 90:925–929
Luine VN, Fischette CT (1982) Inhibition of lordosis behavior by intrahypothalamic implants of pargyline. Neuroendocrinology 34:237–244
Marrone BL, Rodriguez-Sierra JF, Feder HH (1977) Lordosis: inhibiting effects of progesterone in female rats. Horm Behav 8:391–402
Marrone BL, Rodriguez-Sierra JF, Feder HH (1979) Intrahypothalamic implants of progesterone inhibit lordosis behavior in ovariectomized, estrogen-treated rats. Neuroendocrinology 28:92–102
McClintock MK, Adler NT (1978) The role of the female during copulation in wild and domestic Norway rats (Rattus norvegicus). Behavior 67:67–96
Moreines J, Powers JB (1977) Effects of acute ovariectomy on the lordosis response of female rats. Physiol Behav 19:277–283
Morin LP (1977) Progesterone: inhibition of rodent sexual behavior. Physiol Behav 18:701–715
Morin LP, Feder HH (1973) Multiple progesterone injections and the duration of estrus in ovariectomized guinea pigs. Physiol Behav 11:861–865
Morin LP, Feder HH (1974a) Hypothalamic progesterone implants and facilitation of lordosis behavior in estrogen-primed ovariectomized guinea pigs. Brain Res 70:81–93
Morin LP, Feder HH (1974b) Inhibition of lordosis behavior in ovariectomized guinea pigs by mesencephalic implants of progesterone. Brain Res 70:71–80
Myers HI, Young WC, Dempsey EW (1936) Graafian follicle development throughout the reproductive cycle in the guinea pig, with special reference to changes during oestrus (sexual receptivity). Anat Rec 65:381–401
Nadler RD (1970) A biphasic influence of progesterone on sexual receptivity of spayed female rats. Physiol Behav 5:831–835
Neill JD (1980) Neuroendocrine regulation of prolactin secretion. In: Martini L, Ganong WF (eds) Frontiers in neuroendocrinology, vol 6. Raven Press, New York, pp 129–155
Nequin LG, Alvarez JA, Schwartz NB (1979) Measurement of serum steroid and gonadotropin levels and uterine and ovarian variables throughout the 4-day and 5-day estrous cycle of the rat. Biol Reprod 20:659–670
Nock B, Feder HH (1981) Neurotransmitter modulation of steroid action in target cells that mediate reproduction and reproductive behavior. Neurosci Biobehav Rev 5:437–447
Nock B, Blaustein JD, Feder HH (1981) Changes in noradrenergic transmission alters the concentration of cytoplasmic progestin receptors in hypothalamus. Brain Res 207:371–396
Numan M (1974) Medial preoptic area and maternal behavior in the female rat. J Comp Physiol Psychol 87:746–769
Parsons B, Rainbow TC, McEwen BS, Pfaff DW (1981) Oestradiol, sexual receptivity and cytosol progestin receptors in rat hypothalamus. Nature 292:58–59
Parsons B, McEwen BS, Pfaff DW (1982) A discontinuous schedule of estradiol treatment is sufficient to activate progesterone-facilitated sexual behavior and to increase cytosol receptors for progestins in the hypothalamus of the rat. Endocrinology 110:613–619
Perryman RL, Thorner MO (1981) The effect of hyperprolactinemia on sexual and reproductive function in man. J Androl 5:233–242
Pfaff DW (1973) Luteinizing hormone releasing hormone potentiates lordosis behavior in hypophysectomized and ovariectomized female rats. Science 182:1148–1149
Pfaff DW, McEwen BS (1983) Actions of estrogens and progestins on nerve cells. Science 219:807–814
Pfaff DW, Montgomery M, Lewis C (1977) Somatosensory determinants of lordosis in female rats: behavioral definitions of the estrogen effect. J Comp Physiol Psychol 91:134–145
Powers JB (1970) Hormonal control of sexual receptivity during the estrous cycle of the rat. Physiol Behav 5:831–835
Powers JB (1972) Facilitation of lordosis in ovariectomized rats by intracerebral progesterone implants. Brain Res 48:311–325

Powers JB, Moreines J (1976) Progesterone: examination of its postulated inhibitory actions on lordosis during the rat estrous cycle. Physiol Behav 17:493–498
Powers JB, Zucker I (1969) Sexual receptivity in pregnant and pseudopregnant rats. Endocrinology 84:820–827
Psychoyos A (1976) Hormonal control of uterine receptivity for nidation. J Reprod Fertil [Suppl] 25:17–25
Rainbow TC, Parsons B, McEwen BS (1982) Sex differences in rat brain oestrogen and progestin receptors. Nature 300:648–649
Reppert SM, Artman HG, Swaminathan S, Fisher DA (1981) Vasopressin exhibits a rhythmic daily pattern in cerebrospinal fluid but not in blood. Science 213:1256–1257
Rodriguez-Sierra JF, Davis GA (1978) Progesterone does not inhibit lordosis through interference with estrogen priming. Life Sci 22:373–378
Rodriguez-Sierra JF, Komisaruk BR (1982) Common hypothalamic sites for activation of sexual receptivity in female rats by LHRH, PGE_2, and progesterone. Neuroendocrinology 35:365–369
Roos J, Roser S, Vidal N, Aron C (1973) Role played by the ovaries and adrenals in early receptivity during the 4-day cycle of the female rat. Horm Behav 4:341–349
Ross J, Claybaugh C, Clemens LG, Gorski RA (1971) Short latency induction of estrous behavior with intracerebral gonadal hormones in ovariectomized rats. Endocrinology 89:32–38
Rubin BS, Barfield RJ (1983) Progesterone in the ventromedial hypothalamus facilitates estrous behavior in ovariectomized, estrogen-primed rats. Endocrinology 113:797–804
Sakuma Y, Pfaff DW (1980) LH-RH in the mesencephalic central grey can potentiate lordosis reflex of female rats. Nature 283:566–567
Sarkar DK, Chiappa SA, Fink G, Sherwood NW (1976) Gonadotrophin releasing hormone surge in pro-oestrous rats. Nature 264:461–463
Shaikh AA, Shaikh SA (1975) Adrenal and ovarian steroid secretion in the rat estrous cycle temporarily related to gonadotropins and steroid levels found in peripheral plasma. Endocrinology 96:37–44
Smith MS, Freeman ME, Neill JD (1975) The control of progesterone secretion during the estrous cycle and early pseudopregnancy in the rat: prolactin, gonadotropin and steroid levels associated with the rescue of the corpus luteum of pseudopregnancy. Endocrinology 96:219–226
Södersten P (1974) Mounting behavior in pregnant and pseudopregnant female rats. Horm Behav 5:345–354
Södersten P (1976) Lordosis behavior in male, female and androgenized female rats. J Endocrinol 70:409–420
Södersten P, Eneroth P (1981 a) Serum levels of oestradiol-17β and progesterone in relation to sexual receptivity in intact and ovariectomized rats. J Endocrinol 89:45–54
Södersten P, Eneroth P (1981 b) Evidence that progesterone does not inhibit the induction of sexual receptivity by oestradiol-17β in the rat. J Endocrinol 89:63–67
Södersten P, Eneroth (1982) Estradiol and progesterone in the control of sexual receptivity in female rats. Scand J Psychol [Suppl] 1:127–132
Södersten P, Hansen S (1977) Effects of oestradiol and progesterone on the induction and duration of sexual receptivity in cyclic female rats. J Endocrinol 74:477–485
Södersten P, Hansen S (1979) Induction of sexual receptivity by oestradiol benzoate in cyclic rats: influence of ovarian secretions before injection of oestradiol benzoate. J Endocrinol 80:389–395
Södersten P, Eneroth P, Hansen S (1981 a) Neuroendocrine control of daily rhythms in rat reproductive behavior. In: Fuxe K, Gustafsson J-A, Wetterberg L (eds) Steroid hormone regulation of the brain. Pergamon, New York, pp 301–315
Södersten P, Eneroth P, Hansen S (1981 b) Induction of sexual receptivity in ovariectomized rats by pulse administration of oestradiol-17β. J Endocrinol 89:55–62
Södersten P, Pettersson A, Eneroth P (1983 a) Pulsatile administration of estradiol-17β cancels sex difference in behavioral estrogen sensitivity. Endrocrinology 112:1883–1885

Södersten P, Henning M, Melin P, Lundin S (1983 b) Vasopressin alters female sexual behaviour by acting on the brain independently of alterations in blood pressure. Nature 301:608–610
Södersten P, Hansen S, Eneroth P (1983 c) Evidence that prolactin does not affect the induction of sexual behavior by oestradiol and progesterone in ovariectomized rats. J Endocrinol 99:181–187
Södersten P, Hansen S, Eneroth P (1983 d) Inhibition of sexual behavior in lactating rats. J Endocrinol 99:189–197
Tapper CM, Greig F, Brown-Grant K (1974) Effects of steroid hormones on gonadotrophin secretion in female rats after ovariectomy during the oestrous cycle. J Endocrinol 62:511–525
Tartellin MF, Gorski RA (1973) The effects of ovarian steroids on food and water intake and body weight in the female rat. Acta Endocrinol 72:551–568
Tennent BJ, Smith ER, Davidson JM (1980) The effects of estrogen and progesterone on female rat proceptive behavior. Horm Behav 14:65–75
Tennent BJ, Smith ER, Davidson JM (1982) Effects of progesterone implants in the habenula and midbrain on proceptive and receptive behavior in the female rat. Horm Behav 16:352–363
Tobet SA, Baum MJ (1982) Implanation of dihydrotestosterone propionate into the lateral septum inhibits sexual receptivity in estrogen-primed ovariectomized rats. Neuroendocrinology 34:333–338
Wade GN (1976) Sex hormones, regulatory behaviors, and body weight. In: Rosenblatt JS, Hinde RA, Shaw E, Beer CG (eds) Advances in the study of behavior, vol 6. Academic, New York, pp 201–279
Wallen K, Thornton JE (1979) Progesterone and duration of heat in estrogen-treated ovariectomized guinea pigs. Physiol Behav 22:95–97
Wallen K, Goy RW, Phoenix CH (1975) Inhibitory actions of progesterone on hormonal induction of estrus in female guinea pigs. Horm Behav 6:127–138
Whalen RE (1974) Estrogen-progesterone induction of mating in female rats. Horm Behav 5:157–162
Whalen RE, Nakayama K (1965) Induction of oestrous behavior: facilitation by repeated hormone treatments. J Endocrinol 33:525–526
Yanase M, Gorski RA (1976 a) Sites of estrogen and progesterone facilitation of lordosis behavior in the spayed rat. Biol Reprod 15:536–543
Yanase M, Gorski RA (1976 b) The ability of the intracerebral exposure to progesterone on consecutive days to facilitate lordosis behavior: an interaction between progesterone and estrogen. Biol Reprod 15:544–560
Ydstebo R, Södersten P (1977) Induction of sexual receptivity in ovariectomized female rats by subcutaneous implants of estradiol-17β. Horm Behav 9:130–140
Young WC (1961) The hormones and mating behavior. In: Young WC (ed) Sex and internal secretions, vol 2. Williams and Wilkins, Baltimore, pp 1173–1239
Young WC (1969) Psychobiology of sexual behavior in the guinea pig. In: Lehrman DS, Hinde RA, Shaw E (eds) Advances in the study of behavior, vol 2. Academic, New York, pp 1–110
Zemlan FP, Adler NT (1977) Hormonal control of female sexual behavior in the rat. Horm Behav 9:345–357
Zucker I (1966) Facilitatory and inhibitory effects of progesterone on sexual responses of spayed guinea pigs. J Com Physiol Psychol 62:376–381
Zucker I (1967 a) Progesterone in the experimental control of the behavioral sex cycle in the female rat. J Endocrinol 38:269–277
Zucker I (1967 b) Actions of progesterone in the control of sexual receptivity of the spayed female rat. J Comp Physiol Psychol 63:313–316
Zucker I (1968) Biphasic effects of progesterone on sexual receptivity in the spayed guinea pig. J Comp Physiol Psychol 65:472–478
Zucker I, Goy RW (1968) Sexual receptivity in the guinea pig: Inhibitory and facilitatory actions of progesterone and related compounds. J Comp Physiol Psychol 64:378–383

Behavioral Effects on Humans of Progesterone-Related Compounds During Development and in the Adult

STEPHANIE A. SANDERS, M. S. and JUNE MACHOVER REINISCH, Ph. D.[1]

Contents

1 Introduction

Gonadal steroids play an influential role in human development by affecting the expression of behavior and emotions. There are two important periods of hormonal action. The first occurs prenatally. The relative presence or absence of androgenizing hormones, starting at approximately 8 weeks postconception, masculinizes or feminizes the central nervous system, which predisposes the organism – whether genetically male or female – to exhibit higher frequencies of masculine or feminine behavior, respectively (Reinisch 1974, 1983). The second important era of hormonal impact occurs at puberty when the gonads, which have been quiescent during childhood, begin to produce the sex hormones which physically transform the child into a sexually mature individual. The hormonal milieu present prenatally and at puberty influences not only the individual's response to the

1 Kinsey Institute for Research in Sex, Gender, and Reproduction and Department of Psychology, Indiana University, Bloomington, Indiana 47405/USA

Current Topics in Neuroendocrinology, Vol 5

environment but also the response of the social environment to the individual (Reinisch and Gandelman 1978; Reinisch and Sanders to be published). Following sexual differentiation, an individual's current hormonal status is likely to affect responsiveness to various stimuli, thus influencing behavior and demeanor.

The complexity of endocrinological interactions often makes it difficult to evaluate the role of a single hormone in behavioral development. Most models that have been formulated focus on the effects of testosterone or estrogen rather than progesterone. From a psychobiological perspective, the development of a theoretical model to explain the influence of progesterone on behavior is a challenging prospect. Progestogens have both antiandrogenic (Berlin and Meinecke 1981; Kinkcl and Maqueo 1965; Money 1970) and antiestrogenic (Linkie 1982; Klaiber et al. 1982) potential. Yet progesterone also facilitates the expression of certain estrogen-mediated behaviors (Pfaff and McEwen 1983). Several synthetic progestins have been shown to have androgenic effects on development (Reinisch 1983). The fact that synergistic and antagonistic effects may be dose-dependent and that each of the synthetics seems to vary with respect to its relative potency (Tausk 1975), severely hinders interpretation of data, particularly when derived by a variety of methods.

Nevertheless, we have attempted to review the available experimental and quasi-experimental studies relevant to the question of the influence of progesterone-related compounds on human behavior. First, studies are presented which evaluate hormonal effects during the prenatal critical period of sexual differentiation. These are separated into two categories: the effect of exogenous progestogen administration during gestation on long-term behavioral development; and the correlation of umbilical cord blood hormonal concentrations with subsequent behavioral development. Following this is a discussion of studies which examine the effects of progesterone on adult behavior. These are also divided into two groups: progestins as antilibidinal agents; and the influence of the menstrual cycle and oral contraceptives on behavior, affect, and somatic symptom reporting.

2 Effects of Prenatal Administration

Substantial evidence has accrued indicating that prenatal exposure to progestins influences behavioral development in humans. It is not possible, however, to categorize this influence as being generally masculinizing or feminizing. For example, exposure to certain synthetic progestins during critical developmental periods has masculinized the genitalia and behavioral patterns of female animal and human offspring. Yet some of these same compounds have been shown to feminize or demasculinize exposed males. The exact nature of the effect is probably related to the differential chemical configuration of the compound and the dosage, duration, and timing of exposure (Reinisch 1983). Endogenous progesterone of placental or fetal adrenal origin may have a protective effect against androgenization in developing female fetuses (Resko 1974a; Shapiro et al. 1976). Progesterone levels have been found to be significantly higher in female rhesus monkey fetuses than in males (Hagemenas and Kittinger 1972; McDonald et al. 1973;

Resko 1974a, 1975). The higher levels of progesterone available to female fetuses may serve as antiandrogens during developmental periods when the fetus is particularly susceptible to the virilizing action of gonadal or adrenal androgens.

Until recently most clinicians did not distinguish between synthetic progestins and progesterone in terms of pharmacologic effect on treatment outcome. During the past 15 years, however, data have accumulated regarding the possible differential effects of various progestogens administered during gestation. The data on natural progesterone, and progesterone-based and androgen-based synthetic progestins, are reviewed below.

2.1 Progesterone

Dalton (1968a) reported that children whose mothers were treated with progesterone for toxemia of pregnancy were significantly advanced with respect to certain developmental milestones and school performance as compared to untreated controls. In the preliminary study, mothers evidencing pre-eclamptic toxemia were randomly assigned to the progesterone or control group between the 16th and the 28th week postconception.

At 1 year of age, progesterone-exposed offspring were compared to control infants. Although no differences were found between groups "in the time of eruption of the first tooth, total number of teeth, and talking ability", progesterone-exposed infants were significantly advanced in their ability to stand and walk unaided.

In the same report, Dalton examined teacher's achievement ratings of a second sample of children exposed prenatally to progesterone, as compared to two control groups comprised of infants born of untreated toxemic pregnancies and of normal pregnancies and births. When the participants were 9–10 years old, their head teachers were asked to rate them as "above", "average", or "below" standard in five subject areas. Progesterone-exposed students received significantly more "above average" ratings than did either of the comparison groups. The progesterone effect was found to be related to both dosage and timing of administration. Children exposed to 8 g or more of the hormone were rated significantly higher than those who received lower dosages, and those children exposed prior to the 16th week postconception were given significantly higher ratings than those treated later in gestation.

It is relevant to note that the conclusions of this study, the selection of comparison groups, and the analysis of the data have been criticized (Lynch and Mychalkiw 1978). Reanalysis of the data revealed (Lynch and Mychalkiw 1978) that although a significantly greater proportion of progesterone-exposed children stood unaided at 1 year of age as compared to controls, the difference in those walking unaided was not significant. Further, with respect to academic achievement, there were no overall differences in the grades received by exposed and unexposed children. When early-treated offspring were compared to toxemic controls, progesterone-exposed subjects were found to have significantly more "above average" grades in arithmetic. Comparisons to nontoxemic controls were not significant. Evaluations including only high dosage subjects did not produce more significant results.

Two follow-up studies supporting Dalton's (1968a) findings have been presented (Zussman et al. 1975; Dalton 1976). In the first (Zussman et al. 1975), children exposed to progesterone demonstrated a clear advantage in numerical ability and borderline positive effects in spatial and mechanical abilities compared to controls. This effect of progesterone was related to higher dosage and to treatment lasting longer than 8 weeks. Prenatal exposure to progesterone did not seem to affect "masculinity-femininity" scores on the California Personality Inventory or the Bem Sex-Role Inventory.

Retrospective interviews on behavior when subjects were between 5 and 10 years old revealed that those female subjects who had been exposed to larger, longer, and earlier doses of progesterone were less likely to report being a tomboy and had fewer disciplinary reports in school than controls. Progesterone-exposed girls also reported more concern with their appearance. Early exposure was associated with preference for long rather than short hair. Higher and longer dosages were negatively related to influence over peers, especially for females. For males, dosage, duration, and early administration were positively associated with disciplinary reports in school. Progesterone-exposed males also reported less daydreaming about girls, less dating, less time spent thinking about marriage and family life, and decreased incidence of going steady. Exposed children, particularly males, reported playing in larger groups. There was a negative relationship between the dosage and duration of progesterone exposure and the amount of reported outdoor, athletic, and motile types of play in both sexes.

Zussman and colleagues formulated two main conclusions based on these data. The first was that progesterone-exposed subjects were "more studious" and "less socially interactive" than matched controls. They suggested that this hypothesis is a more parsimonious and complete explanation for the results than one based on selective masculinization and femininization in different areas. The second conclusion was that progesterone has an influence on brain development, particularly with longer duration, higher dosage, and early administration.

The second follow-up study (Dalton 1976) compared the educational attainment, at 17–20 years, of prenatally progesterone-exposed offspring to normal and toxemic control subjects. Progesterone-exposed subjects demonstrated a higher level of scholastic achievement. Specifically, a larger proportion graduated with the highest ($p<0.02$) or second highest ($p<0.05$) categories of scores and a greater number went on to university level training. Thirty-two percent of the progesterone subjects obtained admission to a university, as compared to 6.3% of the controls and 6.1% for London in general. "The best academic results were in those whose mothers had received over 5 grams of prenatal progesterone, and for whom administration commenced before the sixteenth week and treatment lasted longer than eight weeks" (Dalton 1976). This is particularly impressive since these progesterone-exposed subjects might actually be expected to perform less well in school due to the fact that their mothers presented symptoms of preeclamptic toxemia. Offspring of toxemic pregnancies exhibit poorer school achievement (Barker and Edwards 1967; Dalton 1968a). Dalton estimated that 25% of these mothers would have developed a "full blown toxemia" if left untreated. According to Dalton (1976), "The progesterone given to the mother not only prevents the development of toxaemia, and eliminates the diminished intelligence in the child but actually enhances the intelligence, the extent of the enhancement being dependent on the time, dosage and duration of treatment". Reevaluation of these data (Lynch and Mychalkiw 1978) confirmed the reported findings (Dalton 1976) but questioned the conclusions. It was suggested that the apparent effect of progesterone might have been an artifact due to the overrepresentation of males in the hormone-exposed sample as compared to that of controls. In England, proportionally more males achieve university level scores, and among those attaining minimum entrance requirements, proportionally more males go on to university studies.

A third investigation lends some additional support to the hypothesis that progesterone may affect school achievement and success (Lynch et al. 1978). Compared to unexposed controls, progesterone-exposed females were found to be significantly more introverted on the Eysenck Personality Inventory. That is, they were characterized as more "quiet and retiring, introspective and fond of books rather than people," liking a "well-ordered life" and rarely behaving "aggressively." Despite the methodological limitations of all these studies, the consistency of the findings suggests some effects of progesterone on personality or temperament, which in turn may enhance school achievement and success.

2.2 Progesterone-Based and Androgen-Based Synthetic Progestins

The offspring of mothers treated with synthetic androgen based progestins during pregnancy were first described in 1967 (Ehrhardt and Money 1967). Personality and IQ were examined in girls born with masculinized genitalia. In-depth interviews conducted with both the subjects and their mothers revealed a higher than expected degree of "tomboyism" in exposed subjects. Tomboyism was defined by a high degree of athletic energy, outdoor pursuits, play with boys' toys and a minimal interest in feminine clothing, dolls, child care, and household chores. Although increased tomboyism seemed to be associated with progestin exposure in prepubertal and high school assessments, a follow-up interview study, which also did not include a comparison group, revealed no apparent long-term effects on career pursuits, social development, or sexual orientation and eroticism in adulthood (Money and Mathews 1982). These findings are reminiscent of those for experimentally androgenized rhesus monkeys, in which clear effects on juvenile behavior were attenuated in adulthood (Goy and Phoenix 1971).

Support for the hypothesis that exogenous progestins administered during pregnancy have long-term effects on the behavioral development of the offspring was provided by studies of children exposed to combinations of androgen-based and progesterone-based synthetic progestins in comparison to their unexposed siblings (Reinisch 1977; Reinisch and Karrow 1977). The matched-sibling comparison design provides partial control for genetic and environmental variables. Birth order was balanced between exposed and unexposed subjects. Genital masculinization was not noted for any of the females. As measured by a standard personality questionnaire, progestin-exposed subjects were significantly more "independent," "individualistic," "selfassured," and "self-sufficient" than their unexposed siblings. Higher scores on these dimensions are more characteristic of males than females. Three of the personality variables affected by progestin exposure have a positive association with scholastic achievement. This suggests that although IQ may not be elevated, scholastic achievement may still be positively affected by prenatal progestin exposure through its influence on personality or temperament (Reinisch 1977).

Utilizing a measure of potential for aggressive response to hypothetical conflict situations, females and males who were prenatally exposed to synthetic progestins were compared to same-sex unexposed siblings (Reinisch 1981). Exposed subjects of both sexes showed a significantly higher potential for physical aggression than their sex-matched siblings. Exposure to androgen-based preparations seemed to be most closely associated with aggressive response. No differences in verbal aggression were found. This difference in physical aggression is consonant with known effects of early exposure to androgen on mammalian behavior. Aggression appears to be one of the most robust of the behavioral sex differences, found across many mammalian species, including humans (Maccoby and Jacklin 1974).

Preliminary findings have been reported regarding the long-term psychological effects of medroxyprogesterone acetate (MPA), a progesterone-based synthetic progestin. Both males (Meyer-Bahlburg et al. 1977) and females (Ehrhardt et al. 1977) from the U.S. who were members of the Buffalo sample of the Collaborative Perinatal Cohort were evaluated. Controls selected from the cohort were matched for sex, race, birth-date (+/− 6 months), socio-economic status, and an attempt was made to match for vaginal bleeding during pregnancy. There were no statistically significant differences between MPA-exposed males and their matched controls. However, the authors reported tendencies for MPA-exposed boys to be less intense in their energy expenditure, and less interested in marriage and having children. Girls who were exposed to MPA in utero demonstrated a significantly stronger preference for attractive, stylish clothes; while most of the unexposed subjects reported either no preference or a clear preference for jeans, pants, and other unisex clothing. There was also a statistically nonsignificant tendency for MPA girls to be less frequently labeled tomboys on a long-term basis.

These findings are consistent in part with those of an earlier report (Zussman et al. 1975) on prenatal progesterone exposure, namely: a lower incidence of tomboyism and a stronger preference for feminine clothing. MPA may enhance traditional patterns of femininity (Ehrhardt et al. 1977). The fact that relatively high dosages of progesterone were found to have subtle but significant demasculinizing effects on some behaviors in genetic males (Zussman et al. 1975) and that statistically nonsignificant tendencies in the same direction were found with exposure to MPA (Meyer-Bahlburg et al. 1977) suggests that the effects of MPA might have been greater if subjects had been exposed to larger doses for longer durations. It should be noted, however, that subjects exposed to generally higher dosages of MPA in combination with other progestins appeared to be masculinized in several dimensions of personality and aggression (Reinisch 1977, 1981; see above). The differences in terms of amount of statistical significance and directionality of these MPA findings as compared to those of our laboratory may derive from the lower dosage and shorter duration treatment regimens of MPA to which these subjects were exposed. Additional evaluations of larger groups of individuals prenatally exposed to progestins are necessary to clarify the role of dosage, timing, and duration of exposure, as well as the differences in the potential effects of the various synthetic progestational compounds.

2.3 Summary

Exposure to exogenous progestins during gestation may have long-term behavioral sequelae. The exact nature of the effect is likely to be dependent upon one or more of the following parameters: (1) the type of preparation; (2) the dosage, timing, and duration of exposure; (3) the nature of the genetic substratum and relative tissue sensitivity. The evidence indicates that progesterone and certain synthetic progestins may enhance scholastic achievement either through effects on specific cognitive abilities or by altering personality/temperament factors which may be associated with achievement motivation. Some synthetic progestins, particularly androgen-based compounds, seem to have masculinizing and/or defeminizing potential in offspring of both sexes. On the other hand, progesterone and lower doses of MPA may have a feminizing and/or demasculinizing effect. Careful consideration of treatment parameters is essential to the interpretation of the increasing number of findings which suggest that prenatal exposure to exogenous progestogens organizes or sensitizes the central nervous system, resulting in long-term behavioral consequences for humans. General statements regarding the developmental effects of a given compound in terms of the amount and direction of masculinization/defeminization or femininization/demasculinization may not be possible because various neurological systems experience their peak sensitivity to gonadal steroids at different points in ontogeny.

3 Perinatal Endogenous Hormones

Most of the research examining the influence of early hormone exposure on human behavioral development has utilized samples of patients with various clinical endocrine disorders, or subjects exposed in utero to exogenous hormones which were administered to their mothers for medical reasons (see above). Another avenue of insight into human behavioral endocrinology is through evaluation of the hormonal content of umbilical cord blood. Such studies allow subsequent assessments of psychological variables to be analyzed in light of the measurements of the hormone concentrations which were actually present at delivery. One advantage of this method is that it provides information about the effects of "normal" levels of *endogenous* hormones, albeit at a unique moment in development. Though the critical period for genital differentiation occurs much earlier in gestation (Abramovich 1974), the critical period(s) for other aspects of physical and behavioral sexual dimorphism might extend throughout gestation. Certainly neurological plasticity is quite high at the time of birth and it is possible that the presence of hormones during this time may influence differentiation of the nervous system.

Differences in the hormonal profiles of the umbilical cord blood from male and female neonates have been noted by a few researchers. Such hormonal differences may be meaningful for later gender development. Studies of both humans (Forest et al. 1974; Maccoby et al. 1979) and rhesus monkeys (Resko 1974b) revealed that males have higher levels of circulating testosterone at birth than do females. Estrogen concentrations seem to be similar in both sexes, though an effect of birth order has been noted (Maccoby et al. 1979; Shutt et al. 1974; Smith 1975).

The findings on sex differences in progesterone concentrations are somewhat contradictory. Laatikainen and Peltonen (1974) reported that boys had higher levels of progesterone; whereas Forest and Cathiard (1978) found 17-α-hydroxyprogesterone (a derivative of progesterone) to be higher in the cord blood of girls. Maccoby et al. (1979, described in detail below) noted no overall sex differences in umbilical progesterone at birth. In parallel studies of rhesus monkeys, some researchers demonstrated that progesterone levels were higher in the fetal circulation of females (Hagemenas and Kittinger 1972; Resko 1975; Resko et al. 1975), while others reported an elevation in males (McDonald et al. 1973). It has been suggested, based on differences between arterial and venous blood, that male fetuses metabolize proportionately less of the available progesterone than do female fetuses (Hagemenus and Kittinger 1972). If sex differences in umbilical cord progesterone do exist, they are probably not robust (Maccoby et al. 1979) and are unlikely to be important in any explanation of behavioral differences between the sexes.

Most research on the effects of prenatal and perinatal hormones upon behavior development has focused on sexual differences. Relatively little attention has been given to the possible relationship between hormones and behavioral differences related to the ordinal position of the child within the sibship. The general assumption has been that sibling-order effects were due solely to such environmental factors as parental experience and the presence or absence of older siblings (Sutton-Smith and Rosenberg 1970), and not to biological variables.

Maccoby et al. (1979) suggested a posteriori that variations in the hormonal milieu at or prior to birth might play a role in the differential development of first born and later-born children. Testosterone, androstenedione, estrone, estradiol, and progesterone concentrations in samples of umbilical cord blood from 256 infants were analyzed by sex and birth order. As mentioned above, testosterone was found in higher concentrations in male infants than in female infants, but no significant sex differences were found for the other four hormones. Additionally, later children who were born less than 4 years after their older sibling were found to have lower concentrations of hormones. The firstborns of both sexes had significantly higher concentrations of progesterone and estrogens, with progesterone showing the largest birth-order effects ($p < 0.001$). The higher progesterone levels of firstborns were not attributable to length of labor, birth weight, or maternal age. For male infants, progesterone values increased as the interval between siblings increased. A lapse of 4 years or more since the birth of the previous child yielded measures of progesterone similar to or higher than that of firstborns. Later-born females also had lower progesterone levels than firstborns; although in contrast to males, the hormone concentrations in females did not seem to be a function of the temporal spacing between deliveries.

The parallel between these hormonal findings and behavioral differences associated with the order and temporal spacing of births suggests a biological component to what has been traditionally viewed as an environmental effect of sibling position. Maccoby et al. (1979) suggested that if a relationship between prenatal exposure to progestogens and enhanced intellectual ability and scholastic achievement exists (Dalton 1968a, 1976; Ehrhardt and Money 1967; Money and Ehrhardt 1972), then the effect of birth order on progesterone levels at delivery may have strong implications for behavioral development. They cite the work of Zajonc and Markus (1975) which demonstrated that the tested intelligence of firstborns is higher than that of later-born children, and that longer intersibling spacing cancelled or even reversed the effect. Maccoby et al. highlight the parallel between these findings and the hypothesized role of progesterone enhancement of intellectual ability. Others, however, have failed to find a birth-spacing effect on overall ability (Belmont et al. 1978). Yet differences in personality and/or temperament which may be related to achievement and specific cognitive abilities have been noted (Koch 1955; McGurk and Lewis 1972; Rosenberg and Sutton-Smith 1969; Sutton-Smith and Rosenberg 1970). Reinisch (1977) did not find differences in IQ between progestin-exposed children and their unexposed siblings, although there was a suggestion from her data that scholastic achievement and success might be enhanced.

Most recently, Jacklin et al. (1983) reported on a subset of the infants (84 boys and 78 girls) which was studied longitudinally. Timidity was assessed in 6- to 18-month-old children by presentation of novel toys. For male children, significant negative relationships were observed between timidity and cord concentrations of progesterone ($p<0.01$) and testosterone ($p<0.02$); whereas estradiol was positively related to timidity ($p<0.1$). No significant correlations were found between hormones and timidity for girls. The results for males are consonant with those of Reinisch (1977) and Reinisch and Karrow (1977), who found differential outcomes for progestin-exposed and estrogen-exposed offspring. Further, progestin-exposed children were reported to be more "independent," "self-assured," and "self-sufficient" than sibling controls (Reinisch 1977; Reinisch and Karrow 1977), and low timidity may be related to these traits.

In summary, the above findings provide tentative albeit indirect support for the existence of a positive relationship between progesterone and enhanced cognitive abilities and/or scholastic achievement. Further, endogenous progesterone may affect the development of temperament or personality. Higher progesterone levels at birth were related to lower timidity in male infants. The lack of correlation between progesterone and timidity in females may indicate the necessity of a male hormonal or genetic substratum for progesterone to have this impact. Future investigations assessing the hormonal content of umbilical cord serum promise to contribute greatly to the understanding of early endogenous hormonal influences on behavioral development.

4 Progesterone in Adulthood

Progestins not only have organizing and/or sensitizing potential in early life, as discussed above, but also demonstrate an activating influence in adulthood when they exist concurrently with the behaviors exhibited. Insight into the influence of progesterone and synthetic progestins on adult human behavior is provided by studies of (1) the antilibidinal effects of progestogens and (2) the behavioral, affective, and somatic correlates of menstrual cyclicity, including examination of oral contraceptive effects.

4.1 Progestogens as Antilibidinal Agents

Exogenous progestins and endogenous progesterone have both been shown to be associated with decreased sexual drive in men and women. The mechanism by which progestogens exert this effect may include both a direct pharmacological action upon neural tissue and an indirect influence by reduction of testosterone and/or estrogen. Data pertaining to the antilibidinal potential of progestogens derive from: (1) studies of the use of medroxyprogesterone acetate in the treatment of male sex offenders; and 2) research into patterns of sexual desire and behavior as they relate to the menstrual cycle and use of oral contraceptives.

4.1.1 MPA Therapy for Male Sex Offenders

One of the most successful therapeutic interventions for the treatment of chronic sex offenders is the administration of cyproterone acetate (used in Europe) or medroxyprogesterone acetate (MPA; used in the U.S.), both of which effectively reduce sex drive (Barry and Ciccone 1975; Gagne 1981; Money 1970; Money et al. 1975, 1976; Pinta 1978; Spodak et al. 1978; Walker 1976). Paraphiliacs, including voyeurs, exhibitionists, erotic sadists, and pedophiliacs, have been successfully treated. MPA reduces testosterone and may diminish obsessive sexual preoccupation and compulsive acting-out behavior, thus facilitating self-control (Berlin and Meinecke 1981). In addition, MPA's efficacy as an antiandrogen has been demonstrated by its effectiveness in treating intractable aggressiveness unrelated to sexual functioning (Blumer and Migeon 1975).

Efforts have been made to titrate the dosage (one or two intramuscular injections per week) so as not to cause total impotence while still maintaining control over the deviant behavior. These effects of MPA appear to be fully reversible within a few months of cessation of treatment. Important for patient acceptance of treatment is the fact that, although strongly antiandrogenic, MPA does not feminize appearance (Physicians' Desk Reference 1981).

Psychiatric counseling is usually a necessary adjunct to MPA therapy (Gagne 1981; Money 1981). In one study, 40 of 48 subjects responded positively within 3 weeks to a combination of MPA and milieu therapy that lasted up to 12 months. Subjects demonstrated lower frequencies of erotic fantasies and arousal, diminished desire for deviant sexual behavior, greater control of sexual urges, and improvement in psychosocial functioning (Gagne 1981). Long-term effects of such treatment even after the termination of medication have been noted and seem to be related to the desire of the patient to control his sexually deviant behavior.

There is evidence that MPA has two modes of action affecting sexual drive. It not only reduces testosterone levels, but also seems to influence the brain directly. Support for this direct effect is provided by the following findings: (1) increasing dosages appear to decrease erotic fantasies even when concentrations of serum testosterone are constant; (2) no compensatory elevation of FSH or LH occurs in response to lower testicular production of testosterone (Berlin and Meinecke 1981).

4.1.2 Sex and the Menstrual Cycle

Some evidence suggests that progestogens may serve as libidinal suppressants in women as well as men. The loss of sexual desire and depression that occur in some women using oral contraceptives may be directly related to the progestin content and inversely related to the estrogen content of the preparation they use (Grant and Davies 1968). Experimentally, progesterone administration has been shown to reduce sexual activity in ovariectomized, estrogen-primed rhesus monkeys (Michael et al. 1967).

Further, cyclic fluctuation in sexual interest and activity in women indicates that endogenous progesterone may be involved in suppression of libidinal drive. In a classic study examining psychological correlates of the menstrual cycle, Benedek and Rubenstein (1939a, b) correlated thematic material from psychoanalytic sessions with menstrual cycle phase as determined by vaginal smears and basal body temperature charts. It was concluded that estrogen activity was associated with increased heterosexual drive, whereas progesterone activity was correlated with a "passive receptive and narcissistic attitude." Some data suggest that both humans (Udry and Morris 1968; Morris and Udry 1982; Williams and Williams 1982) and rhesus monkeys (Michael 1968) demonstrate the lowest rate of sexual activity during the luteal phase, when progesterone levels are highest. To date, the role of gonadal hormones in the sexual behavior of women is uncertain. It is speculated that sexual drive is positively correlated with estrogen and/or androgen and negatively correlated with progesterone (Williams and Williams 1982). The mechanism by which progesterone may inhibit libidinal urges could be related to its antiandrogenic and antiestrogenic properties, to its sedative effect, or to some change in affect, cognition, or perception mediated by the presence of progestin.

4.2 The Menstrual Cycle

Numerous studies have reported that women experience behavioral, affective, and physical changes secondary to the hormonal effects of the menstrual cycle (Dan et al. 1980; Friedman 1982; Komnenich 1981). A brief review of the literature on the psychological effects of menstrual cyclicity, particularly as it relates to progesterone, is germane to the topic of this paper. The reader is advised, however, that the majority of these studies are correlative and that such data are difficult to interpret. As with all studies in behavioral endocrinology, variation in the level of one hormone is related to changes in others. In fact, the interplay between estrogen and progesterone is a hallmark of the menstrual cycle. Another source of error is the presumption of many menstrual researchers that cycles are 28 days in length, and that arbitrary divisions into phases using calendar dates rather than some more accurate index of actual hormonal status is sufficient to make hormone-behavior attributions. An additional problem derives from the dominant cultural belief that menstruation affects behavior and mood negatively (Smith 1975; Weideger 1977; Steiner and Carroll 1977). This stereotype may bias reporting of symptoms as it relates to menstrual cycle phases.

Prior to discussing the psychological correlates of the menstrual cycle, a brief review of the hormonal events of the normal menstrual cycle is in order. During the follicular phase, a gradual rise in estrogen occurs. Just prior to ovulation there is a sharp rise and fall of serum estradiol levels which coincides with the LH peak. This is followed by a secondary, more gradual increase in blood estradiol, which corresponds to the luteal rise in serum progesterone. Levels of both hormones drop precipitously during the premenstruum, prior to the onset of menstruation. Basal body temperature provides a fairly good reflection of the postovulatory rise in progesterone, with the peak temperatures corresponding to the highest progesterone levels (De Allende 1956), demonstrating the hyperthermic effects of progesterone. The thermal nadir is contemporaneous with the estradiol peak which immediately precedes ovulation.

Studies that have focused on the relationship between psychological variables and the menstrual cycle will be examined in an attempt to reveal the potential role of progesterone in activating behavioral response in adulthood. Evidence will be presented in the following manner: (1) studies of symptom reporting by phase of cycle; (2) information regarding the role of progesterone in the etiology of dysmenorrhea and premenstrual syndrome; (3) research on affective and affect-related behavioral changes associated with menstrual cycle phases; (4) studies of specific cognitive changes during the menstrual cycle.

4.2.1 Self-Reports on Menstrual Symptoms

Rudolf Moos (Moos 1968a) developed what is now the most widely used standardized questionnaire on menstrual distress – the Menstrual Distress Questionnaire (MDQ). The MDQ asks subjects to rate themselves on 47 symptoms using a six-point scale ranging from "no experience" of a symptom to "acute or partially disabling experience" of that symptom. Subjects rate themselves on the symptoms for their "most recent flow," the "week before," and the "remainder of the cycle," thus allowing for comparison of menstrual, premenstrual, and intermenstrual symptom reporting. Factor analysis of the items yielded eight symptom scales: pain (cramps, general aches and pains), changes in concentration (lower judgment, distractible), behavior change (lower school or work performance, decreased efficiency), autonomic reactions (dizziness, faintness, nausea, vomiting), water retention (bloating, breast tenderness, weight gain), negative affect (restlessness, tension, depression, irritability), arousal (excitement, feeling of well-being, affectionate), control (feeling of suffocation, chest pains, palpitations), and an additional item regarding change in eating habits. The pain scale includes symptoms usually associated with dysmenorrhea, whereas the dimensions of negative affect and water retention reflect symptoms traditionally associated with premenstrual syndrome.

In a sample of 839 women (Moos 1968a, 1977) the factors of pain, concentration, behavior change, and autonomic reactions were found to have higher mean scores menstrually. Water retention and negative affect were higher premenstrually. With the exception of arousal and control, all factors showed large differences when intermenstrual scores were compared to either premenstrual or menstrual responses. Approximately 30% of the women respondents complained of pain, behavior change, water retention, and negative affect. Between 10% and 15% reported symptoms found on the concentration and autonomic reaction scales. These menstrual cycle phase effects have been replicated by a number of independent researchers (Golub 1976; Gruba and Rorhbaugh 1975; Silbergeld et al. 1971; Sommer 1973). Further, the daily self-report form of the MDQ has confirmed the original findings (Moos et al. 1969; Moos 1977; Silbergeld et al. 1971).

There are large individual differences in symptom reporting both within and across menstrual cycle phases. When individual profiles were compiled by transforming scale scores so that each had a standardized mean and standard deviation, women presented a wide variety of symptom patterns. It was found that some women reported phase effects with their worst symptoms occurring premenstrually, others rated their menstrual symptoms as the worst, a third group recorded both their premenstrual and menstrual symptoms as being equally bad, and a fourth reported no cyclic changes (Moos 1977). In one sample, a surprisingly large subgroup of 8% reported higher arousal (a set of positive reactions) menstrually and premenstrually while not complaining of any of the negative symptoms (Moos 1977). It appears from these findings that the most promising approach to understanding the etiology of menstrual symptomology lies in distinguishing various symptom subgroups not only by the pattern of complaints but also by the timing of the symptoms within the cycle.

Some insight into the contribution of hormonal factors to symptom reporting might be provided by comparison of menstrual distress reports by women who are taking oral contraceptives with those who are not. In Moos' initial normative sample there were 420 women currently taking oral contraceptives and 298 who were not (Moos 1977). Greater severity of symptoms both menstrually ($p<0.01$ for 10 of 47 symptoms) and premenstrually ($p<0.01$ for 15 of 47 symptoms) was reported by the women not on oral contraceptives. Most of the symptoms fell into the areas of pain, concentration, negative affect, and behavior change. In contrast, other investigations have found no differences in symptom reporting between those women taking birth control pills and those who were not (Doty 1975, cited in Moos 1977; Stultz 1971 reported in Moos 1977). Based on unsolicited comments made by women in the original MDQ sample, Moos reported that approximately 10% of the women who had taken oral contraceptives (either currently or in the past) specifically stated that they had far less menstrual distress during pill usage. Of this same group, another 10% said that they had worse symptoms while taking oral contraceptives. Clearly, there is substantial variablity among women in response to exogenous hormone administration.

A comparison of symptom reports between women on sequential pills and those on combination preparations indicated that sequential preparations were associated with significantly more premenstrual complaints on the water retention and negative affect scales (Moos 1968b, 1977). This suggests that cyclically fluctuating hormone levels (perhaps progesterone concentrations in particular), whether endogenous or exogenous in origin, are more likely to be associated with premenstrual complaints than are relatively tonic states. The fact that premenstrual and menstrual distress are more likely to occur in ovulatory rather than anovulatory cycles (Benson 1980; Green 1977) also supports this hypothesis.

One of the basic problems with using any self-report symptom questionnaire such as the MDQ is that the reports are likely to reflect cultural stereotypes rather than the actual experience of the women (Brooks et al. 1977; Parlee 1974; Ruble 1977). To hypothesize regarding the role of progesterone in the cyclicity of menstrual symptoms based on data from nonclinical samples is extremely difficult, due to variability among women, the general lack of physiological indicators of menstrual status, and the confounding of subjects' expectations for effects with subjective self-ratings. However, the data are suggestive of at least a contributory role of progesterone in the development of perimenstrual distress.

4.2.2 Role of Progesterone in Premenstrual Syndrome and Dysmenorrhea

4.2.2.1 Premenstrual Syndrome

Frank (1931) labelled the irritability, depression, and tension that occur in some women just prior to and/or during menstruation as premenstrual tension (PMT). More recently, the cyclic premenstrual recurrence of symptoms (usually within 5 days of flow onset) is known as premenstrual syndrome (PMS). As many as 150 symptoms have been identified as part of this syndrome. The physical symptoms most often include breast tenderness and swelling, abdominal bloating and discomfort, and headaches. Psychologically, increased irritability, depression, and tension are most frequently cited. The number of women specifically reporting premenstrual symptoms ranges from approximately 25% to 90% (Coppen and Kessel 1963; Janiger et al. 1972; Moos 1968a, 1969; Shaines 1961). However, probably only 5%–10% are severely affected by PMS (Norris and Sullivan 1983). Age seems to be related to menstrual distress, with younger women experiencing more dysmenorrhea (Green 1977) and older women having more premenstrual complaints (Gough 1975; Moos 1977).

The etiology of PMS is believed by many to involve an absolute or relative deficiency of progesterone, which directly or indirectly induces the symptoms typical of this syndrome (Backstrom and Cartensen 1974; Backstrom and Mattson 1975; Brush 1977, 1979; Dalton 1979; Munday 1977). A number of hypotheses regarding the exact etiology and mechanism involved in PMS have been proposed but will not be reviewed here (see reviews, e.g. Abraham 1980; Rausch and Janowsky 1982). Recently, public attention has focused on PMS due to (1) its successful use as a mitigating factor in the defense of criminal cases in the United Kingdom (Dalton 1980) and France (Gonzales 1981) and (2) the fact that PMS sufferers are demanding treatment including progesterone administration as suggested by Dalton (1977). This demand has resulted in the establishment of a number of clinics and support groups throughout the United States.

Dalton (1964, 1977) enthusiastically supports the use of progesterone in the treatment of PMS, despite the absence of controlled studies to support its efficacy. She reports great success with such treatment. However, the two double-blind studies in the literature have failed to demonstrate any greater therapeutic response to progesterone as compared to placebo. Intramuscular progesterone treatment failed to reduce premenstrual depression in one group of women (Smith 1976). In the second study, which utilized a double-blind crossover design, Sampson (1979) found no significant difference in therapeutic effectiveness between progesterone and placebo. In fact, placebo was more effective than progesterone in the majority of cases. Recently, dydrogesterone treatment has been prescribed for some patients (Taylor and James 1979), with improvement noted in 70% of those treated when a single-blind placebo-controlled design was employed (Day 1979). In this latter study, 40% of subjects in the control condition responded to placebo. The therapeutic value of dydrogesterone may be due to pharmacological effects rather than compensation for progesterone deficiency, since women with low and normal levels of progesterone respond similarly (Brush 1979).

Dalton (1977) has contended that synthetic progestins would not be effective in treating PMS because they would decrease already low progesterone levels while not providing the protective physiological action of the natural hormone (e.g., natriuresis, glucocorticoid transport). Others support the use of synthetic progestins based upon their belief that fluctuation of hormones rather than absolute levels is an important etiological factor in PMS (Johansson 1971; Studd 1979). According to this view, oral contraceptives which attenuate the oscillation of endogenous ovarian steroids should provide protection and/or relief from PMS symptoms. Large survey studies of long-term pill users generally find that premenstrual complaints, including water retention and negative affect (irritability, depression, and mood swings), are reduced (Herzberg and Coppen 1970; Kutner and Brown 1972; Moos 1977; Rouse 1978; Wood et al. 1979). In short-term double-blind investigations, results have been contradictory (Green 1982). Sibergeld et al. (1971) found that cyclic variability in aggression, anxiety, and irritability were reduced by oral contraceptives, while the premenstrual somatic symptoms (nausea, water retention, breast tenderness) were worsened and extended across the cycle. Cullberg (1972) reported that hormone-treated groups showed significantly more adverse mental changes throughout the cycle when compared to placebo control subjects. One problem with studies involving short-term use of oral contraceptives is that negative side effects of the pill are most likely to occur during the period of initial use and tend to subside within a few months (Green 1982). Administration of progestins without conjugated estrogens has been tested for therapeutic effectiveness in patients manifesting PMS. Twenty-seven patients with moderate to severe premenstrual symptoms were found to be less irritable when given norethisterone or a diuretic as compared to placebo (Coppen et al. 1972). On the other hand, medroxyprogesterone plus a diuretic was found to be no more successful in relieving premenstrual complaints than was placebo (Jordheim 1972).

In summary, the effectiveness of progesterone administration for the treatment of PMS is questionable. Oral contraceptives may alleviate premenstrual symptoms after an initial period of adjustment to the regular ingestion of exogenous hormones. Perhaps a contributing factor to our lack of understanding of PMS is the simplistic view of it as a single entity. As the work of Abraham (1978, 1980) suggests, it is very likely that symptom subgroups exist, reflecting different etiologies requiring differential treatment.

4.2.2.2 Dysmenorrhea

The term dysmenorrhea refers to painful menstruation, which is the most common of all gynecological complaints and is the leading cause of absenteeism according to Benson (1980). Endogenous prostaglandins play a significant role in the etiology of primary dysmenorrhea. Dysmenorrheic patients have higher levels of prostaglandins (PGE_2 and $PGF_2\alpha$) in the endometrium, endometrial fluid, and menstrual fluid than do nondysmenorrheic women (Best and Pickles 1963; Bydgeman 1964; Chan and Jennifer 1978; Embrey 1975; Lundstrom and Green 1978; Pickles et al. 1965). Progesterone plays an important stimulatory role in the production of prostaglandins during the secretory phase of the menstrual cycle (Fuchs 1982). Estrogen is also necessary since all progesterone effects in the uterus are dependent on estrogen-priming. The importance of progesterone in this process is exemplified by the fact that in anovulatory cycles the endometrium has been shown to contain only 20% of the amount of prostaglandins found on corresponding days of ovulatory cycles (Fuchs 1982). The exact mechanism by which ovarian steroids influence the synthesis of prostaglandins remains unclear.

4.2.3 The Menstrual Cycle and Affect

In light of the prevalent view that women are more susceptible to depression, anxiety, and irritability in the perimenstruum than at other times, the lack of consistency of findings among studies examining affective correlates of the menstrual cycle is remarkable. Various research methodologies have been applied in attempts to clarify the relationship between hormones and mood or affect-related behaviors.

In some investigations anxiety, hostility, and depression have been shown to increase premenstrually (Moos et al. 1969; Golub 1976). Depression and anxiety as measured by the Depression Adjective Checklist (DACL) and the State-Trait Anxiety Inventory, respectively, were found to increase significantly during the premenstrual phase in women over 30 years old (Golub 1976), but showed no cyclic effects in adolescent girls (Golub and Harrington 1981). Based upon data derived using projective techniques, anxiety levels at ovulation were significantly lower than they were premenstrually (Ivey and Bardwick 1968). Another study found that negative affect decreased between days 4 and 10 of the cycle and then increased between days 16 and 26 (when progesterone levels are high) (Paige 1971).

Others have failed to find a relationship between the hormonal fluctuations of the menstrual cycle and mood (Abplanalp et al. 1979; Persky 1974). Premenstrual decline in mood was identified, using retrospective reports in one sample, only when the subjects were aware of the purpose of the research. When the intent of the investigation was disguised, no premenstrual increase in negative mood was found (Vila and Beech 1980). In addition, fluctuations in mood were more likely to be associated with stressful events of daily living than with menstrual cycle phase, when daily measurement of affect was obtained (Wilcoxon et al. 1976).

A few studies have specifically examined the relationship between endogenous estrogen and progesterone levels and mood. No relationship was demonstrated between plasma levels of ovarian hormones and measures of daily mood (POMS) and the enjoyment of activities (Abplanalp et al. 1979). Blood samples were drawn three times a week from 14 women, except during the ovulatory phase when four to five samples per week were taken. There was no systematic variation of either positive or negative affect across the five endocrinologically defined phases of the cycle. Thus, it was concluded that life events are more influential in determining affective states than is the hormonal milieu. Supporting this finding, no significant correlations were obtained between ovarian hormone levels and negative affect as reported on the MDQ (Doty et al. 1981). Comparison of high and low symptom reporters yielded no significant differences in mean plasma estradiol and progesterone, or patterns of hormone change across the menstrual cycle (Taylor 1979).

The relationship between affect and the introduction of exogenous hormonal stimulation is difficult to interpret across studies. An examination of the composition of oral contraceptive preparations suggested that progestins produced depression, whereas estrogens stimulated anxiety and nervous tension (Cullberg 1972). An interview study (Bardwick 1976) concluded that the ingestion of progestin-dominant contraceptives was correlated with increased deference, nurturance, and affiliation. Estrogen-dominant pills were associated with increased assertiveness, aggression, and hostility. Although these findings are not contradictory they fail to provide a coherent picture of the different effects of exogenous progestins versus exogenous estrogens.

Finally, with regard to emotions and hormonal status, a relationship between psychiatric disturbance and the menstrual cycle has been identified. Relying upon postmortem examinations, researchers have reported that suicide victims were significantly more likely to be menstruating than not (MacKinnon and McKinnon 1956; Ribeiro 1962). Dalton (1959), analyzing admission records of two large mental hospitals in England, found cycle-related peaks in admissions for acute psychiatric disturbance, attempted suicide, and depressive illness. In each category the highest incidence occurred during the first four days of menstruation, the secondary peak was during the four days preceding onset of flow, and the third highest rate was found at midcycle. These findings were corroborated by a study of calls to the Los Angeles Suicide Prevention Center (Mandell and Mandell 1967). The role of gonadal hormones in severe depression is not clear. However, if a women is subject to severe depression perhaps associated with PMS, declining levels of progesterone may contribute to development of suicidal ideation over several perimenstrual days.

In general, studies of the affective correlates of ovarian hormone cyclicity are inconsistent and inconclusive. Perhaps higher depression, more negative affect, and decreased sexual interest are produced by higher progesterone levels, while higher estrogen levels facilitate sexual activity and assertiveness. These effects may be mediated by expectations of increased negative affect prior to and during menstruation. The apparent discrepancies among studies examining the influence of endogenous and exogenous hormones on cyclic variation in mood may be a reflection of the importance of relative ratios of estrogen to progesterone, the rise and fall of steroid levels, or the presence or absence of other factors which remain obscure. Further, contradictory findings may be attributable in part to the failure to distinguish between those women who suffer from PMS or dysmenorrhea and those who do not. Future investigation in this area will benefit from the technological advances in terms of hormonal assays, and current knowledge which suggests that: (1) It is important to divide women into symptom subgroups, thus allowing for the possibility that different menstrual symptom patterns may have distinct etiologies; (2) individual variability in responsiveness to gonadal hormones must be considered in all investigations of behavioral endocrinology; and (3) consideration must be given to social factors and to the individual's expectation of menstrual debilitation as they relate to cyclic variation in affect and symptom reporting.

4.2.4 Criminality and the Menstrual Cycle

Positive correlations have also been noted between the premenstrual and menstrual phases of the cycle and the commission of violent crimes (Cooke 1945; Dalton K. 1961, 1980; d'Orban and Dalton J. 1980). It has been suggested that increased criminality, psychiatric disturbances, and accidents are related to the occurrence and severity of premenstrual tension symptoms (Dalton K. 1964, 1979). Examination of the offenses of the 156 women studied by K. Dalton (1961) reveals that the "crimes of violence" included only theft, soliciting, and drunkenness. Labelling these as crimes of violence in association with the menstrual cycle is misleading, especially in light of the fact that alcohol tolerance is lower during the premenstruum (Jones and Jones 1976).

The prevailing view that premenstrual symptoms are associated with an increased likelihood of commission of violent crime has been questioned. D'Orban and J. Dalton (1980) studied the records of 50 women prisoners convicted of violent crimes (most involving some form of assault or criminal property damage). Confirming the positive relationship between cycle phase and criminal behavior, a significantly higher proportion (44%, $n=22$, $p<0.02$) of the women were reported to have committed crimes during the perimenstrual period (days 1–4, 25–28). It is important to note, however, that a significant lack of offenses ($p<0.01$) occurred during the ovulatory and post-ovulatory phases of the cycle (days 13–24), suggesting that ovulation may be negatively associated with crime. Further, only two of the women in this study claimed that there was any connection between the commission of the crime and their menstrual cycle. There was no tendency for women who report premenstrual symptoms to commit their crimes during the perimenstruum. Thus, any connection between PMS and criminality is at best questionable. Caution must be exercised in the interpretation of the results of the d'Orban and J. Dalton study because of the reliance on retrospective self-report for the dates of the last menstrual periods (LMP). Of the 19 cases (out of 50) for which prison records also listed LMP, only seven were corroborated. Even if it were true that criminals are more likely to have committed their offenses during the premenstrual and menstrual phases of the cycle, it would not necessarily be true that women in general who are in the premenstrual and menstrual phases of their cycles are more likely to commit crimes (Parlee 1973).

4.2.5 Changes in Cognitive Ability Across the Menstrual Cycle

4.2.5.1 General Cognitive Functioning

The most frequently cited research examining intellectual performance and the menstrual cycle is two studies by Katharina Dalton (1960, 1968 b). The results of both of these studies support the idea of a perimenstrual performance decrement. The first study reported a premenstrual/menstrual (5 days for each phase) decline in the examination scores of English schoolgirls aged 11–17 years. Dalton found that perimenstrual scores were lower for 27%, higher for 17%, and demonstrated no differences for 56% of the subjects. Evaluation of these findings is difficult since data were not statistically analyzed.

The second study (Dalton 1968 b) reported relationships between menstrual cycle phases and scores obtained by adolescent schoolgirls on advanced (A) and ordinary (O) level standardized examinations. The findings were as follows for the 16–19-year-olds taking the A levels: (1) Girls in the premenstrual and menstrual phases of the cycle obtained 3% lower average grades than those who were intermenstrual at the time of the exams; (2) there was a 9% lower distinction rate for girls in the premenstrual phase than in the intermenstrual phase; and (3) the pass rate was 13% lower for the premenstrual group than for the intermenstrual group. A similar pattern was obtained by reviewing O level performance: a 5% decline in scores in the premenstrual as compared to the intermenstrual phase. Again, no statistical tests of significance were applied to the findings. It was concluded that a female student taking an examination during the premenstrual or menstrual phase of her cycle was subject to an average 5% handicap. These findings might be used to support the notion that perimenstrual performance decrements are causally related to declining serum progesterone levels.

The validity of these data has been challenged by a number of researchers (see Parlee 1973, 1982; Sommer 1973, 1982). A major criticism is that the performance of a given individual was not compared across her own menstrual cycle, but rather the scores obtained by different girls at different phases were compared. In the second study (Dalton 1968 b), 342 papers were submitted by the 125 girls. Thus, there were not equal numbers of grades for all subjects. This inequality put disproportionate weight on those individuals who turned in more papers within any given cycle (Sommer 1982). Further, cycle length was reported by 42% of the girls to be longer or shorter than usual, with a higher proportion than expected menstruating within the examination week. In light of this fact, it seems reasonable to suggest that the apparent correlation between test performance and phase of the menstrual cycle might be attributable to another variable, such as anxiety or stress, which elicited both poorer test performance and menstruation (Sommer 1982).

This explanation seems probable, especially considering the failure of other studies to support Dalton's conclusions regarding overall intellectual performance (Bernstein 1977; Eagleson 1927; Lough 1937; Seward 1934; Sommer 1972), despite a publication bias toward reporting only positive results (Sommer 1982). For example, evaluations of the performance of approximately 200 college women on weekly administrations of the Watson-Glaser Critical Thinking Appraisal test and biweekly class examinations revealed no significant relationships between phase of the menstrual cycle and performance (Sommer 1972). The investigators collected menstrual cycle data after testing had been completed, so that subjects were blind to the purposes of the study. In addition, both within-subject and across-subject comparisons were made, separating oral contraceptive users from nonusers. Employing a similar paradigm and controlling for scholastic ability and motivation, investigators found no performance deficit when scores from the perimenstrual period (4 days prior to and following the onset of menstrual flow) were compared to intermenstrual scores (Bernstein 1977). "In summary, higher level intellectual function in normal healthy women appears to be stable and independent of menstrual fluctuation" (Sommer 1982).

4.2.5.2 Specific Cognitive Functioning

Although there seems to be no evidence of *overall* cognitive impairment perimenstrually, a relationship may exist between specific abilities and hormone levels. When specific cognitive or perceptual-motor tasks are examined, mixed results are found. Broverman et al. (1981), based upon their own research in electrophysiology, depression, and cognitive style as they relate to ovarian hormones, have suggested a way of interpreting the complex data generated by different laboratories. Two categories of cognitive style exist – automatization and perceptual-restructuring – which may be differentially affected by the relative amounts of estrogen and progesterone. Automatized behaviors are those which require little conscious attention because they are so highly practiced. According to Broverman et al. (1981), tasks which measure automatization ability, such as speed of reading, color naming, and naming objects, are likely to be facilitated in the high-estrogen preovulatory phase of the menstrual cycle and impeded by the elevated progesterone levels in the luteal phase. For example, speech, which is an automatized behavior, seems to be affected by menstrual cycle phase (Silverman and Zimmer 1975, 1976). An increase in speech disfluencies during the luteal and menstrual phases was found when tape recordings were analyzed across the menstrual cycle (Komnenich 1974). In contrast, perceptual-restructuring tasks require inhibition of the initial "automatized" response to the obvious properties of a stimulus array, in order to appropriately respond to the hidden or embedded stimulus characteristics. Examples of such tasks include the Embedded Figures test and backward counting. According to Broverman et al., tasks which demand perceptual-restructuring skills are likely to improve in the high progesterone post-ovulatory phase of the cycle. This hypothesis was supported in a study of 87 regularly menstruating women (Broverman et al. 1981).

Relationships between several cognitive tests and hormonal state were evaluated, using serum assays to determine levels of estrogen, progesterone, testosterone, follicle-stimulating hormone, and luteinizing hormone (Komnenich 1978). Comparisons were made among women – using and not using oral contraceptives – and men in tests demanding either automatization or perceptual-restructuring. For women, the menstrual cycle was divided into four phases based upon information from the hormone assays and calendar dates, while male subjects were arbitrarily assigned phase designations. Supporting the Broverman et al. hypothesis, "the highest performance on simple repetitive tasks and the lowest performance on tasks requiring inhibition occurred in the ovulatory phase" (Komnenich et al. 1978). Data from birth control pill users and men showed no such effect. High estrogen levels were credited with the decline in perceptual-restructuring performance. There was no relationship, however, between absolute serum concentrations of estrogen and performance on these measures. That is, subjects with higher overall levels of estrogen did not do less well. The progestins contained in the oral contraceptives may have provided protection from the performance decrement at midcycle either directly, or indirectly by suppressing ovulation or attenuating the fluctuations of hormonal levels across the menstrual cycle. In another investigation, enhanced performance on the Embedded Figures test and Human Figure Drawing test were noted during the 3rd week (high progesterone) of the cycle for two groups of women – 116 girls and unmarried women, and 155 college students (Dor-Shav 1976). Finally, simple calculations were found to be solved more quickly and more accurately in the luteal phase than during follicular days in 16 normally menstruating young women (Wuttke et al. 1976). Optimum performance was achieved 2–3 days prior to the onset of menstrual flow. These findings provide support for the hypothesis that endogenous progesterone plays a role in specific cognitive functioning.

The differential effects of three oral contraceptives with varying progestogen content were evaluated with regard to affect, automatization, and perceptual-restructuring ability (Worsley 1980). Women were given Minovlar (containing the weakest progestin), Eugynon (containing both biologically active and inactive isomers of the progestin), or Neogynon (containing probably the most antiestrogenic progestin). All preparations contained equal dosages of estrogen. Each woman was tested prior to administration of the medication and then again within 3 months of the initial use of the contraceptive. Statistical analyses discriminated Neogynon from the other preparations. This preparation was found to be associated with increased negative affect and decreased automatization performance. Therefore, either the greater estrogenic activity of Eugynon and Minovlar or the greater progestogenic effect of Neogynon may explain the lower automatization ability related to ingestion of Neogynon. The results support the hypothesis that progestogens, perhaps in conjunction with estrogens, affect behavioral expression.

Some investigations have failed to find correlations between specific task performance and hormonal variation. Fifty women ranging in age from 20 to 45 years were given 13 cognitive tests assessing speed and fluency of thought, memory, and reasoning, and two measures of affect (Golub 1976). No significant performance differences were found between the premenstrual and intermenstrual phases of the cycle. Further, no reliable correlations between cognitive performance and premenstrual measures of mood were found. A second study with essentially negative findings evaluated 14 women on a number of cognitive and perceptual tasks (Zimmerman and Parlee 1973). The only phase difference in performance that was noted was an improvement in arm-hand steadiness during the luteal phase over that in the premenstrual phase. In spite of these contradictory findings, the weight of currently available evidence suggests an influence of both endogenous and exogenous progestins on specific aspects of cognitive function.

4.2.6 Electrophysiological and Neurophysiological Correlates of Hormonal Stimulation

The effect of estrogen and progesterone on central nervous system functioning has been examined in relation to differential cognitive styles across the menstrual cycle (Klaiber et al. 1982). Specifically, EEG photic driving responses and differential performance on cognitive tasks pre- and postovulation have been studied. A relationship between these variables and the ability of estrogen to enhance central adrenergic functioning, in particular the action of norepinephrine, has been suggested.

The term "EEG photic driving response" is used to describe the tendency of EEG alpha waves to synchronize with the frequency of a bright flashing light. This alpha driving response is affected by substances that stimulate or block norepinephrine. Drugs which interfere with adrenergic functioning, such as the phenothiazines, reportedly increase the frequency of EEG photic driving responses (Jorgensen and Wulff 1958; Wilson and Glotfelthy 1958; Misurec and Nahanek 1964; Killam et al. 1967). Conversely, agents which enhance adrenergic activity such as norepinephrine, amphetamine, nicotine (cigarette smoking), and monoamine oxidase (MAO) inhibitors were found to diminish the frequency of EEG photic driving responses (Floru et al. 1962; Shetty 1971; Vogel et al. 1974, 1977). This alpha driving response has been used as an indicator of adrenergic functioning within the central nervous system.

Changes in EEG driving responses have been noted across the menstrual cycle. Postovulatory increases in the frequency of EEG photic driving responses were found in healthy women who were studied for two ovulatory menstrual cycles as determined by basal body temperature records (Vogel et al. 1971). In addition, the effects of exogenous hormone administration on the alpha driving responses in amenorrheic women were examined. Administration of oral conjugated estrogens resulted in a significant decrease in frequency of the EEG driving responses, suggesting enhancement of central adrenergic states. However, administration of medroxyprogesterone in conjunction with estrogen yielded frequencies of EEG photic driving responses significantly higher than those recorded for estrogen alone, indicating a depression of central adrenergic function. Thus, support for the hypothesis that estrogen and progesterone affect EEG photic driving response has come from both correlative and manipulative studies (Klaiber et al. 1982).

The mode of action of these gonadal hormones on EEG alpha driving responses is most likely due to their effects on norepinephrine. Estrogen administration has been shown to potentiate the activity of norepinephrine through the following mechanisms: (1) increasing the synthesis of norepinephrine (Greengrass and Tonge 1974; Kendall and Narayana 1978); (2) decreasing the rate of degradation of catecholamines by enzyme inhibition (Ball et al. 1972; Luine et al. 1977; Srivastava et al. 1978; Klaiber et al. 1971, 1972; Crowley et al. 1978); and (3) blocking synaptic reuptake of norepinephrine (Janowsky and Davis 1970; Inaba and Kamata 1979).

Progesterone and testosterone may block or oppose the neurophysiological effects of estrogen (Klaiber et al. 1982; Rodier 1971). Apparently progesterone decreases the activity of norepinephrine in three ways: (1) by counteracting estrogenic enhancement of activity of the rate-limiting enzyme in the synthesis of norepinephrine (Beattie et al. 1973); (2) by inhibiting estrogen's ability to maintain levels of norepinephrine (Crowley et al. 1978); and (3) by reversing the estrogen-induced inhibition of the norepinephrine-degrading enzyme MAO (Klaiber et al. 1971). It has been suggested (Klaiber et al. 1982) that central nervous system adrenergic functioning (particularly norepinephrine) is enhanced by the relatively high estrogen to progesterone ratio in the preovulatory phase of the menstrual cycle as compared with the postovulatory phase, in which higher levels of progesterone counteract the effects of estrogen.

Automatized and perceptual-restructuring behaviors are differentially affected by adrenergic stimulants and depressants. Adrenergic stimulants (e. g., caffeine, amphetamine), which increase norepinephrine, enhance performance on automatized behaviors (Callaway and Stone 1960; Florey and Gilbert 1943). Adrenergic depressants (e. g., chlorpromazine), which decrease norepinephrine, enhance performance on perceptual-restructuring tasks (Broverman et al. 1968; Helper et al. 1963; Shatin et al. 1956). Thus, there appears to be a strong relationship among high progesterone, low norepinephrine, and facilitation of perceptual-restructuring ability. Conversely, high estrogen, high norepinephrine, and increased automatization performance are positively related.

In summary, the interconnections between cognitive performance, gonadal hormones, and adrenergic functioning, though complex, provide perhaps the best model to date for understanding at least some of the behavioral effects of the menstrual cycle. More research on the interplay between hormones and neurotransmitters is necessary to elucidate the mechanisms by which endogenous hormones affect behavior.

5 Conclusion

This review represents an attempt to collate pertinent psychological and behavioral data on humans relating to the role of progestins during two of the important eras of hormonal impact – gestation and the reproductive years. Despite deficiencies in the number and scope of studies conducted and occasional inconsistencies in the findings obtained, the consensus appears to be that endogenous and exogenous progestins have the capability of influencing the development and expression of human behavior. Early exposure may organize or sensitize the neural substrate, thus establishing behavioral predispositions. The presence of progestins in adult life may activate particular behaviors. The specific directionality and magnitude of effects are likely to be dependent on the nature of the compound, parameters of exposure (including dosage, timing, and duration), and the sensitivity of the organism to the preparation. Finally, a basic congruence was noted between the effects produced by progestogens as a result of fetal exposure, and stimulation during the reproductive era. That is, prenatal exposure to progestins appears to organize behaviors that enhance school achievement and success, such as specific cognitive abilities and/or dimensions related to affect (temperament and/or personality) which may operate either separately or in concert. Similarly, progestin stimulation during adulthood influences these same behavioral dimensions – affective and cognitive – as reflected in the studies of sexual/aggressive behavior and psychological correlates on the menstrual cycle. Future research must concentrate on discriminating the relative influence of progestins on each of these sets of factors during successive periods of development.

Acknowledgments. This work was supported in part by grants from the National Institute of Child Health and Human Development (HD 17655) to JMR. We would like to thank R. Beasley, J. Brewer, and C. S. Kaufman for editorial comments and L. Berg, D. Frasier, K. Newburg, and A. Smith for bibliographic assistance. We would also like to thank K. A. Lanyi and L. A. Rosenblum for their assistance.

References

Abplanalp JM, Donnelly AF, Rose RM (1979) Psychoendocrinology of the menstrual cycle: I. Enjoyment of daily activities and moods. Psychosom Med 41:587–604

Abplanalp JM, Rose RM, Donnelly AF, Livingston-Vaughan L (1979) Psychoendocrinology of the menstrual cycle: II. The relationship between the enjoyment of activities, mood and reproductive hormones. Psychosom Med 41:605–615

Abraham GE (1978) The normal menstrual cycle. Discussion I: the menstrual cycle. In: Givens JR (ed) Endocrine causes of menstrual disorders, Year Book Medical, Chicago, pp 15–50

Abraham GE (1980) Premenstrual tension. Curr Probl Obstet Gynecol 3(12):1–39

Abramovich DR (1974) Human sexual differentiation – in utero influences. Br J Obstet Gynaecol 81:448–453

Backstrom T, Cartensen H (1974) Estrogen and progesterone in plasma in relation to premenstrual tension. J Steroid Biochem 5:257–260

Backstrom T, Mattsson B (1975) Correlation of symptoms in premenstrual tension to oestrogen and progesterone concentrations in blood plasma: a preliminary study. Neuropsychobiology 1:80–86

Ball P, Knuppen R, Haupt M, Breuer H (1972) Interactions between estrogens and catechol amines. III. Studies on the methylation of catechol estrogens, catechol amines and other catechols by the catechol-o-methyltransferase of human liver. J Clin Endocrinol 34:736–746

Bardwick JM (1976) Psychological correlates of the menstrual cycle and oral contraceptive medication. In: Sacher ES (ed) Hormones, behavior, and psychopathology. Raven, New York, pp 95–105

Barker DJP, Edwards JH (1967) Obstetric complications and school performance. Br Med J 2:695

Barry DJ, Ciccone JR (1975) Use of depo-provera in the treatment of aggressive sexual offenders: preliminary report of three cases. Bull Am Acad Psychiatry Law 3(3):179–184

Beattie CW, Rodgers CH, Soyka LF (1973) Influence of ovariectomy and ovarian steroids on hypothalamic tyrosine hydroxylase activity in the rat. Endocrinology 91:276–286
Belmont L, Stein Z, Zybert P (1978) Child spacing and birth order: effect on intellectual ability in two-child families. Science 202:995–996
Benedek T, Rubenstein BB (1939 a) The correlation between ovarian activity and psychodynamic processes. I. The ovulatory phase. Psychosom Med 1:245–270
Benedek T, Rubenstein BB (1939 b) The correlation between ovarian activity and psychodynamic processes. II. The menstrual phase. Psychosom Med 1:461–485
Benson RC (1980) Current obstetrics and gynecology diagnosis and treatment, 3rd edn. Lange, Los Altos
Berlin FS, Meinecke CF (1981) Treatment of sex offenders with antiandrogenic medication: conceptualization, review of treatment modalities, and preliminary findings. Am J Psychiatry 138(5):601–607
Bernstein BE (1977) Effect of menstruation on academic performance among college women. Arch Sex Behav 4:289–296
Best FA, Pickles VR (1963) A myometrial effect of oestradiol imitated by high Mg++ concentrations. J Physiol 166:12p-13p
Blumer D, Migeon C (1975) Hormone and hormonal agents in the treatment of aggression. J Nerv Ment Dis 160:127–137
Brooks J, Ruble DN, Clarke AE (1977) College women's attitudes and expectations concerning menstrual-related changes. Psychosom Med 39:288–298
Broverman DM, Klaiber EL, Kobayashi Y, Vogel W (1968) Roles of activation and inhibition in sex differences in cognitive abilities. Psychol Rev 75:23–50
Broverman DM, Majcher D, Shea D, Paul V, Klaiber EL, Vogel W (1981) Changes in cognitive task performance across the menstrual cycle. J Comp Physiol Psychol 95:646–654
Brush MG (1977) The possible mechanism causing the premenstrual tension syndrome. Curr Med Res Opin 4(Suppl 4):9
Brush MG (1979) Endocrine and other biochemical factors in the aetiology of the premenstrual syndrome. Curr Med Res Opin 6(Suppl 5):19–27
Bygdeman M (1964) The effect of different prostaglandins on human myometrium in vitro. Acta Physiol Scand [Suppl] 63:242
Callaway E III, Stone G (1960) Re-evaluating focus of attention. In: Uhr L, Miller JC (eds) Drugs and behavior. John Wiley, New York, pp 393–398
Chan WY, Jennifer CH (1978) Determination of menstrual prostaglandin levels in non-dysmenorrheic and dysmenorrheic subjects. Prostaglandins 15:365
Cooke WR (1945) The differential psychology of the American woman. Am J Obstet Gynecol 49:457–472
Coppen A, Kessel N (1963) Menstruation and personality. Br J Psychiatry 109:711–721
Coppen AJ, Milne HB, Outram DH, Weber JCP (1972) Dytide, norethisterone and a placebo in the premenstrual syndrome – a double-blind comparison. Clin Trials J 6:33–36
Crowley WR, O'Donohue TL, Wachslight H, Jacobowitz DM (1978) Effect of estrogens and progesterone on plasma gonadotrophins and on catecholamine levels and turnover in discrete brain regions of ovariectomized rats. Brain Res 154:345–357
Cullberg J (1972) Mood changes and menstrual symptoms with different gestagen/estrogen combinations. Acta Psychiatr Scand [Suppl] 236:3–74
Dalton K (1959) Menstruation and acute psychiatric illness. Br Med J 1:148–149
Dalton K (1960) Effect of menstruation on schoolgirls' weekly work. Br Med J 1:326–328
Dalton K (1961) Menstruation and crime. Br Med J 2:1752–1753
Dalton K (1964) The premenstrual syndrome. Thomas, Springfield
Dalton K (1968 a) Ante-natal progesterone and intelligence. Br J Psychiatry 114:1377–1382
Dalton K (1968 b) Menstruation and examinations. Lancet 2:1386–1388
Dalton K (1976) Prenatal progesterone and educational attainments. Br J Psychiatr 129:438–442
Dalton K (1977) The premenstrual syndrome and progesterone therapy. Heineman, London

Dalton K (1979) Once a month. Hunter House, Pomona, CA
Dalton K (1980) Cyclical criminal acts in premenstrual syndrome. Lancet 2:1070–1071
Dan AJ, Graham EA, Beecher CP (eds) (1980) The menstrual cycle (vol 1): a synthesis of interdisciplinary research. Springer, New York Heidelberg Berlin
Day JB (1979) Clinical trials in the premenstrual syndrome. Curr Med Res Opin 6(Suppl 5):40–45
De Allende ILC (1956) Anovulatory cycles in women. Am J Anat 98:293–305
d'Orban PT, Dalton J (1980) Violent crime and the menstrual cycle. Psychol Med 10:353–359
Dor-Shav NK (1976) In search of premenstrual tension: note on sex differences in psychological differentiation as a function of cyclical physiological changes. Percept Mot Skills 42:1138–1142
Doty RL, Snyder PJ, Huggins GR, Lowry LD (1981) Endocrine, cardiovascular and psychological correlates of olfactory sensitivity changes during the human menstrual cycle. J Comp Physiol Psychol 95(1):45–60
Eagleson HE (1927) Periodic changes in blood pressure, muscular coordination and mental efficiency in women. Comp Psychol Monogr 4:65
Ehrhardt AA, Money J (1967) Progestin-induced hermaphroditism: IQ and psychosexual identity in a study of ten girls. J Sex Res 3:83–100
Ehrhardt AA, Grisanti GC, Meyer-Bahlburg HF (1977) Prenatal exposure to medroxyprogesterone acetate (MPA) in girls. Psychoneuroendocrinology 2:391–398
Embrey MP (1975) The prostaglandins in human reproduction. Churchill Livingstone, New York
Florey CD, Gilbert J (1943) The effects of benzedrine sulfate and caffeine citrate on the efficiency of college students. J Appl Psychol 27:121–134
Floru R, Costin A, Nestianu V, Sterescu-Volanschi M (1962) Research concerning the effect of noradrenaline upon the electrical activity of the central nervous system and upon evoked rhythm of intermittent photic stimulation in cats with chronic electrodes. Electroencephalogr Clin Neurophysiol 14:566
Forest MG, Cathiard AM (1978) Ontogenic study of plasma 17-α-hydroxyprogesterone in the human I: postnatal period: evidence for transient ovarian activity in infancy. Pediatr Res 12:5–11
Forest MG, Sizonenko PC, Cathiard AM, Bertrand J (1974) Hypophysogonadal function in humans during the first year of life. J Clin Invest 53:819–828
Frank RT (1931) The hormonal causes of premenstrual tension. Arch Neurol Psychol 26:1053–1057
Friedman RC (ed) (1982) Behavior and the menstrual cycle. Dekker, New York
Fuchs F (1982) Dysmenorrhea and dyspareunia. In: Friedman RC (ed) Behavior and the menstrual cycle. Dekker, New York, pp 199–216
Gagne P (1981) Treatment of sex offenders with medroxyprogesterone acetate. Am J Psychiatry 138(5):644–646
Golub S (1976) The effect of premenstrual anxiety and depression on cognitive function. J Pers Soc Psychol 34(1):99–104
Golub S, Harrington DM (1981) Premenstrual and menstrual mood changes in adolescent women. J Pers Soc Psychol 41(5):961–965
Gonzales ER (1981) Premenstrual syndrome: an ancient woe deserving of modern scrutiny. JAMA 245(14):1393–1396
Gough H (1975) Personality factors related to reported severity of menstrual distress. J Abnorm Psychol 84:59–65
Goy RW, Phoenix CH (1971) The effects of testosterone propionate administered before birth on the development of behavior in genetic female rhesus monkeys. In: Sawyer CH, Gorski RA (eds) Steroid hormones and brain function. University of California Press, Berkeley
Grant ECG, Davies JP (1968) Effect of oral contraceptives on depressive mood changes and on endometrial monoamine oxidase and phosphates. Br Med J 3:777–780
Green J (1982) Recent trends in the treatment of premenstrual syndrome: a critical review. In: Friedman RC (ed) Behavior and the menstrual cycle. Dekker, New York, pp 367–395

Green TH (1977) Gynecology, 3rd edn. Little, Brown, Boston
Greengrass PM, Tonge SR (1974) The accumulation of noradrenalin and 5-hydroxytryptamine in three regions of mouse brain after tetrabenazine and iproniazid: effects of ethinyloestradiol and progesterone. Psychopharmacologia 39:187–191
Gruba G, Rorhbaugh M (1975) MMPI correlates of menstrual distress. Psychosom Med 37:265–273
Hagemenas FC, Kittinger GW (1972) The influence of fetal sex of plasma progesterone levels. Endocrinology 91:253–256
Helper MM, Wilcott RC, Garfield SL (1963) Effects of chlorpromazine on learning and related processes in emotionally disturbed children. J Consult Clin Psychol 27:1–9
Herzberg B, Coppen A (1970) Changes in psychological symptoms in women taking oral contraceptives. Br J Psychiatry 116:161–164
Inaba M, Kamata K (1979) Effect of estradiol-17 beta and other steroids on noradrenaline and dopamine binding to synaptic membrane fragments of rat brain. J Steroid Biochem 11:1491–1497
Ivey ME, Bardwick JM (1968) Patterns of affective fluctuations in the menstrual cycle. Psychosom Med 30:336–345
Jacklin CN, Maccoby EE, Doering CH (1983) Neonatal sex-steroid hormones and timidity in 6–18-month-old boys and girls. Dev Psychobiol 16(3):163–168
Janiger O, Riffenburgh R, Kersh R (1972) Cross-cultural study of premenstrual symptoms. Psychosomatics 13:226–235
Janowsky DS, Davis JM (1970) Progesterone-estrogen effects on uptake and release of norepinephrine by synaptosomes. Life Sci 9:525–531
Johanssen EDB (1971) Depression of progesterone levels in women treated with synthetic gestogens after ovulation. Acta Endocrinol 68:779–792
Jones BM, Jones MK (1976) Alcohol effects on women during the menstrual cycle. Ann NY Acad Sci 273:576–587
Jordheim O (1972) The premenstrual syndrome-clinical trials of treatment with a progestogen combined with a diuretic compared with both a progestogen alone and with a placebo. Acta Obstet Gynecol Scand 51:77–80
Jorgensen RS, Wulff MH (1958) The effects of orally administered chlorpromazine on the electroencephalogram of man. Electroencephalogr Clin Neurophysiol 10:325
Kendall DA, Narayana K (1978) Effects of oestradiol-17beta on monoamine concentrations in the hypothalamus of anoestrous ewe. J Physiol 282:44–45
Killam KF, Killam EK, Naquet R (1967) An animal model of light sensitive epilepsy. Electroencephalogr Clin Neurophysiol 22:497–513
Kinkel FA, Maqueo M (1965) Prevention of steroid-induced sterility in neonatal male and female rats. Endocrinology 77:859–862
Klaiber EL, Kobayashi Y, Broverman DM, Hall F (1971) Plasma monoamine oxidase activity in regularly menstruating women and in amenorrheic women receiving cyclic treatment with estrogens and progestin. J Clin Endocrinol Metab 33:630–638
Klaiber EL, Broverman DM, Vogel W, Kobayashi Y, Moriarty D (1972) Effects of estrogen therapy on plasma MAO activity and EEG driving responses of depressed women. Am J Psychiatry 128:1492–1498
Klaiber EL, Broverman DM, Vogel W, Kennedy JA, Nadeau CJL (1982) Estrogens and central nervous system function: electroencephalography, cognition, and depression. In: Friedman RC (ed) Behavior and the menstrual cycle. Dekker, New York, pp 267–289
Koch HL (1955) Some personality correlates of sex, sibling position, and sex of sibling among five and six year old children. Genet Psychol Monogr 52:3–50
Komnenich P (1974) Hormonal influences on verbal behavior in women. Diss Abst Int 34:3065B
Komnenich P, Lane DM, Dickey RP, Stone SC (1978) Gonadal hormones and cognitive performance. Physiol Psychol 6:115–120
Komnenich P, McSweeney M, Noack JA, Elder N (eds) (1981) The menstrual cycle (vol 2): Research and implications for women's health. Springer, New York Heidelberg Berlin

Kutner SJ, Brown WL (1972) History of depression as a risk factor for depression with oral contraceptives and discontinuance. J Nerv Ment Dis 155:163–169

Laatikainen T, Peltonen J (1974) Levels of estriol, estriol sulfate, progesterone, and neutral steroid mono- and disulfates in umbilical cord arterial and venous plasma, vol 32. Inserm, Paris

Linkie DM (1982) The physiology of the menstrual cycle. In: Friedman RC (ed) Behavior and the menstrual cycle. Dekker, New York, pp 1–21

Lough OM (1937) A psychological study of functional periodicity. J Comp Psychol 24:359–368

Luine VN, Khylchevskaya RI, McEwen BS (1975) Effects of gonadal steroids on activities of monoamine oxidase and choline acetylase in rat brain. Brain Research 86:293–306

Luine VN, McEwen BS, Black IB (1977) Effect of 17beta-estradiol on hypothalamic tyrosine hydroxylase activity. Brain Res 120:188–192

Lundstrom V, Green K (1978) Endogenous levels of prostaglandin F2alpha and its main metabolites in plasma and endometrium of normal and dysmenorrheic women. Am J Obstet Gynecol 130:640

Lynch A, Mychalkiw W (1978) Prenatal progesterone II: its role in the treatment of preeclamptic toxaemia and its effects on the offspring's intelligence: a reappraisal. Early Hum Dev 2(4)323–339

Lynch A, Mychalkiw W, Hutt SJ (1978) Prenatal progesterone I: its effects on development and on intellectual and academic achievement. Early Hum Dev 2:305–322

Maccoby EE, Jacklin CN (1974) The psychology of sex differences. Stanford University Press, Stanford, CA

Maccoby EE, Doering CH, Jacklin CN, Kraemer H (1979) Concentrations of sex hormones in umbilical-cord blood: their relation to sex and birth order of infants. Child Dev 50:632–642

MacKinnon PCB, MacKinnon IL (1956) Hazards of the menstrual cycle. Br Med J 1:555

Mandell AJ, Mandell MP (1967) Suicide and the menstrual cycle. J Am Med Assoc 200:792–793

McDonald GJ, Yoshinaga K, Greep BO (1973) Progesterone values in *Macaca mulatta* near term. Am J Phys Anthropol 38:201–206

McGurk H, Lewis M (1972) Birth order: a phenomenon in search of an explanation. Dev Psychol 7:366

Meyer-Bahlburg HF, Grisanti GC, Ehrhardt AA (1977) Prenatal effects of sex hormones on human male behavior: medroxyprogesterone acetate (MPA). Psychoneuroendocrinology 2:383–390

Michael RP (1968) Gonadal hormones and the control of primate behavior. In: Michael RP (ed) Endocrinology and human behavior. Oxford University Press, London

Michael RP, Saayman G, Zumpe D (1967) Inhibition of sexual receptivity in rhesus monkeys. J Endocrinol 39:309–310

Misurec J, Nahanek K (1964) Changes in photo-metrazol (photo-myoclonic) threshold in treatment with some psychopharmacological drugs. Electroencephalogr Clin Neurophysiol 17:607

Money J (1970) Use of androgen-depleting hormone in the treatment of sex offenders. J Sex Res 6:165–172

Money J (1981) Paraphilia and abuse-martyrdrom: exhibitionism as a paradigm for reciprocal couple counseling combined with antiandrogen. J Sex Marital Ther 7(2):115–123

Money J, Ehrhardt AA (1972) Man and woman, boy and girl: the differentiation and dimorphism of gender identity from conception to maturity. Johns Hopkins University Press, Baltimore, MD

Money J, Mathews D (1982) Prenatal exposure to virilizing progestins: an adult follow-up study of twelve women. Arch Sex Behav 11(1):73–83

Money J, Wiedeking C, Walker P, Migeon C, Meyer W, Borgaonkar D (1975) 47,XYY and 46,XY males with antisocial and/or sex-offending behavior: antiandrogen therapy plus counseling. Psychoneuroendocrinology 1:165–178

Money J, Wiedeking C, Walker PA, Gain D (1976) Combined antiandrogenic and counseling program for treatment of 46,XY and 47,XYY sex offenders. In: Sachar EJ (ed) Hormones, behavior, and psychopathology. Raven, New York, pp 105–120

Moos R (1968a) The development of a menstrual distress questionnaire. Psychosom Med 30:853–867

Moos R (1968b) Psychological aspects of oral contraceptives. Arch Gen Psychiatry 19:87–94

Moos R (1969) A typology of menstrual cycle symptoms. Am J Obstet Gynecol 103:390–402

Moos RH (1977) Menstrual distress questionnaire manual. Social Ecology Laboratory, Department of Psychiatry, Stanford University, Veterans Administration Hospital, Palo Alto, California

Moos R, Kopell B, Melges F, Yalom I, Lunde D, Clayton R, Hamburg D (1969) Fluctuations in symptoms and moods during the menstrual cycle. J Psychosom Res 13:37–44

Morris NM, Udry JR (1982) Epidemiological patterns of sexual behavior in the menstrual cycle. In: Friedman RC (ed) Behavior and the menstrual cycle. Dekker, New York, pp 129–153

Munday M (1977) Hormone levels in severe premenstrual tension. Curr Med Res Opin 4(Suppl 4):16–22

Norris RV, Sullivan C (1983) PMS: Premenstrual syndrome. Rawson, New York

Paige KE (1971) Effects of oral contraceptives on affective fluctuations associated with the menstrual cycle. Psychosom Med 33:515–537

Parlee MB (1973) The premenstrual syndrome. Psychol Bull 80:454–465

Parlee MB (1974) Stereotypic beliefs about menstruation: a methodological note on the Moos menstrual distress questionnaire and some new data. Psychosom Med 36:229–240

Parlee MB (1982) The psychology of the menstrual cycle: biological and psychological perspectives. In: Friedman RC (ed) Behavior and the menstrual cycle. Dekker, New York, pp 77–99

Persky H (1974) Reproductive hormones, moods, and the menstrual cycle. In: Friedman RC, Richart RM, Vande Wiele RL, Stern LO (eds) Sex differences in behavior. John Wiley, New York, pp 455–466

Pfaff DW, McEwen BS (1983) Actions of estrogens and progestins on nerve cells. Science 219:808–814

Physicians' Desk Reference (1981) 35th edn. Medical Economics, Oradell

Pickles VR, Hall WJ, Best FA, Smith GN (1965) Prostaglandins in endometrium and menstrual fluid from normal and dysmenorrheic subjects. J Obstet Gynaecol Br Common 72:185–192

Pinta ER (1978) Treatment of obsessive homosexual pedophilic fantasies with medroxyprogesterone acetate. Biol Psychiatry 13(3):369–373

Rausch JL, Janowsky DS (1982) Premenstrual tension: etiology. In: Friedman RC (ed) Behavior and the menstrual cycle. Dekker, New York, pp 397–427

Reinisch JM (1974) Fetal hormones, the brain, and human sex differences: a heuristic integrative review of the recent literature. Arch Sex Behav 3:51–90

Reinisch JM (1977) Prenatal exposure of human foetuses to synthetic progestin and oestrogen affects personality. Nature 266:561–562

Reinisch JM (1981) Prenatal exposure to synthetic progestins increases potential for aggression in humans. Nature 211:1171–1173

Reinisch JM (1983) Influence of early exposure to steroid hormones on behavioral development. In: Everaerd W, Hindley CB, Bot A, van der Werff ten Bosch JJ (ed) Development in adolescence: psychological, social, and biological aspects. Nijhoff, The Netherlands, pp 63–113

Reinisch JM, Gandleman R (1978) Human research in behavioral endocrinology: methodological and theoretical considerations. In: Dorner G, Kawakami M (eds) Hormones and brain development. Elsevier, Amsterdam, pp 77–86

Reinisch JM, Karrow WG (1977) Prenatal exposure to synthetic progestins and estrogens: effects on human development. Arch Sex Behav 6:257–288

Reinisch JM, Sanders SA (1983) Hormonal influences on sexual development and behavior. In: Schwartz MF, Moraczewski AS, Monteleone JA (eds) Sex and gender: A theological and scientific inquiry: The Pope John center, St. Louis, Missouri, p. 48–64

Resko JA (1974a) The relationship between fetal hormones and the differentiation of the central nervous system in primates. In: Montagna W, Sadler WA (eds) Reproductive behavior: Plenum, New York, p 211

Resko JA (1974b) Sex steroids in the circulation of the fetal and neonatal rhesus monkey: a comparison between male and female fetuses, vol 32. Inserm, Paris

Resko JA (1975) Fetal hormones and their effects on the differentiation of the central nervous system of primates. Fed Proc 34:1650–1655

Resko JA, Ploem JG, Stadelman HL (1975) Estrogens in fetal and maternal plasma of the rhesus monkey. Endocrinology 97:425–430

Ribeiro AL (1962) Menstruation and crime. Br Med J 1:640

Rodier WI III (1971) Progesterone-estrogen interactions in the control of activity-wheel running in the female rat. J Comp Physiol Psychol 74:365–373

Rosenberg BG, Sutton-Smith B (1969) Sibling age spacing effects on cognition. Dev Psychol 1:661–668

Rouse P (1978) Premenstrual tension: a study using the Moos menstrual distress questionnaire. J Psychosom Res 22:215–222

Ruble DW (1977) Premenstrual symptoms: a reinterpretation. Science 197:291–292

Sampson GA (1979) Premenstrual syndrome: a double-blind controlled trial of progesterone and placebo. Br J Psychiatry 135:209–215

Seward GH (1934) The female sex rhythm. Psychol Bull 31:153–192

Shaines N (1961) A re-evaluation of some aspects of femininity through a study of menstruation: a preliminary report. Comp Psychiatry 2:20–26

Shapiro BH, Goldman AS, Bongiovanni AM, Marison JM (1976) Neonatal progesterone and feminine sexual development. Nature 264:795–796

Shatin L, Rockmore L, Funk IC (1956) Response of psychiatric patients to massive doses of thorazine. II. Psychological test performance and comparative drug evaluation. Psychiatr Q 30:402–416

Shetty T (1971) Photic responses in hyperkinesis of childhood. Science 174:1356–1357

Shutt DA, Smith ID, Shearman RP (1974) Fetal plasma steroids in relation to parturition III: The effect of parity and method of delivery upon umbilical oestrone oestradiol levels. J Obstet Gynaecol Br Common 81:968–970

Silbergeld S, Brast N, Noble E (1971) The menstrual cycle: a double-blind study of symptoms, mood, and behavior, and biochemical variables using Enovid and placebo. Psychosom Med 33:411–428

Silverman EM, Zimmer CH (1975) Speech fluency fluctuations during the menstrual cycle. J Speech Hear Res 18:202–206

Silverman EM, Zimmer CH (1976) Replication of "speech fluency fluctuation during the menstrual cycle". Percept Mot Skills 42:1004–1006

Smith ID, Shutt DA, Shearman RP (1975) Foetal plasma steroid concentrations related to gestational age and method of delivery. J Steroid Biochem 6:657–662

Smith SL (1975) Mood and the menstrual cycle. In: Sacher EJ (ed) Topics in psychoendocrinology. Grune and Stratton, New York

Smith SL (1976) The menstrual cycle and mood disturbance. Clin Obstet Gynecol 19:391–397

Sommer B (1972) Menstrual cycle changes and intellectual performance. Psychosom Med 34:263–269

Sommer B (1973) The effect of menstruation on cognitive and perceptual-motor behavior: a review. Psychosom Med 35:515–534

Sommer B (1982) Cognitive behavior and the menstrual cycle. In: Friedman RC (ed) Behavior and the menstrual cycle. Dekker, New York, pp 101–127

Spodak MK, Falck ZA, Rappeport JR (1978) The hormonal treatment of paraphiliacs with depo-provera. Criminal Justice and Behavior 5(4):304–314
Srivastava K, DasGupta PK, Mehrotra PK (1978) Monoamine oxidase activity in the pituitary of mice after estrogen, progesterone and centchroman treatment. Indian J Physiol Pharmacol 22:310
Steiner M, Carroll BJ (1977) The psychobiology of premenstrual dysphoria: a review of theories and treatments. Psychoneuroendocrinology 2:321–335
Studd J (1979) Premenstrual tension syndrome. Br Med J 1:410
Sutton-Smith B, Rosenberg BG (1970) The sibling. Holt, Rinehart and Winston, New York
Tausk M (1975) Pharmacology of hormones. Thieme, Stuttgart
Taylor JW (1979) Plasma progesterone, oestradiol-17B and premenstrual symptoms. Acta Psychiatr Scand 60(1):76–86
Taylor RW, James CE (1979) The clinician's view of patients with premenstrual syndrome. Curr Med Res Opin 6 (Suppl 5):46–51
Udry JR, Morris NM (1968) Distribution of coitus in the menstrual cycle. Nature 220:593–596
Vila J, Beech HR (1980) Premenstrual symptomatology: an interaction hypothesis. Br J Soc Clin Psychol 19:73–80
Vogel W, Broverman DM, Klaiber EL (1971) EEG responses in regularly menstruating women and in amenorrheic women treated with ovarian hormones. Science 172:388–391
Vogel W, Broverman DM, Klaiber EL, Kobayashi Y (1974) EEG driving responses as a function of monoamine oxidase. Electroencephalogr Clin Neurophysiol 36:205–207
Vogel W, Broverman DM, Klaiber EL (1977) EEG responses to photic stimulation in habitual smokers and nonsmokers. J Comp Physiol Psychol 91:418–422
Walker PA (1976) Medroxyprogesterone acetate: an antiandrogen for the rehabilitation of sex offenders. International Congress on Sexology, Montreal, Canada
Weideger P (1977) Menstruation and menopause: the physiology and psychology, the myth and reality. Dell, New York
Wilcoxon LA, Schrader SL, Sherif CW (1976) Daily self-reports on activities, life events, moods and somatic changes during the menstrual cycle. Psychosom Med 38:399–417
Williams GD, Williams AM (1982) Sexual behavior and the menstrual cycle. In: Friedman RC (ed) Behavior and the menstrual cycle. Dekker, New York, pp 155–176
Wilson WP, Glotfelthy JS (1958) Effect of intravenous promazine on arousal responses in man. Dis Nerv Syst 19:307–309
Wood C, Larsen L, Williams R (1979) Social and psychological factors in relation to premenstrual tension and menstrual pain. Aust NZ J Obstet Gynecol 19:111–115
Worsley A (1980) A prospective study of the effects of the progestagen content of oral contraceptives on measures of affect, automatization, and perceptual restructuring ability. Psychopharmacology 67:289–296
Wuttke W, Arnold P, Becker D, Creutzfeldt O, Langenstein S, Tirsch W (1976) Hormonal profiles and variations of the EEG and of performance in psychological tests in women with spontaneous menstrual cycles and under oral contraceptives. In: Itil TM (ed) Psychotropic action of hormones. Spectrum, Jamaica, NY
Zajonc RB, Markus GB (1975) Birth order and intellectual development. Psychol Rev 82:74–88
Zimmerman E, Parlee MB (1973) Behavioral changes associated with the menstrual cycle: an experimental investigation. J Appl Soc Psychol 3:335–344
Zussman JU, Zussman PP, Dalton K (1975) Post-pubertal effects of prenatal administration of progesterone. Society for Research in Child Development, Denver, Colorado

Subject Index

MIX
Papier aus verantwortungsvollen Quellen
Paper from responsible sources
FSC® C105338

If you have any concerns about our products,
you can contact us on
ProductSafety@springernature.com

In case Publisher is established outside the EU,
the EU authorized representative is:
Springer Nature Customer Service Center GmbH
Europaplatz 3, 69115 Heidelberg, Germany

Printed by Libri Plureos GmbH
in Hamburg, Germany